Martin Hubert Borecki

Pinning and Wetting Models for Polymers with semi-flexible Interaction

Martin Hubert Borecki

Pinning and Wetting Models for Polymers with semi-flexible Interaction

Random Polymer Models

Südwestdeutscher Verlag für Hochschulschriften

Imprint

Any brand names and product names mentioned in this book are subject to trademark, brand or patent protection and are trademarks or registered trademarks of their respective holders. The use of brand names, product names, common names, trade names, product descriptions etc. even without a particular marking in this work is in no way to be construed to mean that such names may be regarded as unrestricted in respect of trademark and brand protection legislation and could thus be used by anyone.

Publisher:
Südwestdeutscher Verlag für Hochschulschriften
is a trademark of
Dodo Books Indian Ocean Ltd., member of the OmniScriptum S.R.L Publishing group
str. A.Russo 15, of. 61, Chisinau-2068, Republic of Moldova Europe
Printed at: see last page
ISBN: 978-3-8381-2006-5

Zugl. / Approved by: Berlin, TU, Diss., 2010

Copyright © Martin Hubert Borecki
Copyright © 2010 Dodo Books Indian Ocean Ltd., member of the OmniScriptum S.R.L Publishing group

Für Sheila
auf ewig

Acknowledgment

Working as a Ph.D. student in Berlin, Zürich and Padova was a magnificent as well as challenging experience to me. During all these years various people contributed directly or indirectly to the successful completion of this work. It was hardly possible for me to thrive in my doctoral work without the precious support of these personalities. Here is a small tribute to all those people.

First and foremost, I would like to thank my supervisor Prof. Jean-Dominique Deuschel for introducing me to an interesting and vibrant research area. I am very much obliged to his valuable guidance and advice throughout all the past years. I am also very indebted to Prof. Francesco Caravenna, who kindly agreed to become the co-examiner of this thesis. His great hospitality and supervision during my pleasant visit in Padova lead to results given in chapter two. Besides, also for encouragement and comments to numerous questions I just have to say: grazie. Moreover, I sincerely thank Prof. Erwin Bolthausen for the stimulating discussions and for having the opportunity of taking part in the stochastic program of the university and ETH during a whole term in Zürich. I really enjoyed the familiar and good working atmosphere there. In addition I thank Dr. Frank Aurzada for general discussions on Gaussian processes and of course all the others around me for finding a sympathetic ear.

Thanks also to all my old and new colleagues at the mathematical departments of the TU, Humboldt Universität, Universität Potsdam, WIAS, Universität Zürich, ETH and Università degli Studi di Padova for having good times together.

I owe my loving thanks to my wife Sheila for her continuous moral support, patience and understanding, especially when I was abroad. I wish to thank my parents and my sister for their constant support and encouragement in all my professional endeavors: dziękuję wam za zaufanie i ciągłą pomoc przez całe moje życie.

I am grateful to Deutsche Forschungsgemeinschaft for providing financial support through the International Research Training Group "Stochastic Models of Complex Processes". It was a pleasure to participate in this great project. Finally, I also thank the Berlin Mathematical School (BMS) for the fellowship in the phase II graduate program.

Preface

This book deals with the stochastic description of models motivated from the point of view of semi-flexible polymers. These are chain-like molecules build up from small molecular units (monomers) and, depending on the length-scale, displaying different flexibility properties. In particular, we are interested in the behavior of such a polymer chain in the proximity of an attractive environment, e.g. a membrane. An important question in this situation is whether the polymer sticks close to the membrane (localization) or fluctuates away from it (delocalization).

We consider a directed model of a polymer chain given by the spatial position of the i-th monomer in the d-dimensional space. The polymer is subject to an interaction with itself (semi-flexibility) as well as an interaction with the environment. The environment has an attractive effect and is represented by an m-dimensional subspace (reference plane). When touching the reference plane the polymer receives an energy-reward in the form of a pinning parameter. The interaction among individual monomers is represented by a mixture of gradient and Laplacian interaction with different interaction potentials. We call this model the pinning model and study particularly the localization behavior of the polymer in proximity of a reference plane. That means, we ask if a phase transition occurs by modifying the pinning parameter. In this case we would have a critical pinning value, such that above this value the polymer chain has a positive contact fraction at the reference plane (localization). On the other hand it would not be the case for a pinning parameter below the critical value (delocalization). This behavior additionally depends on the choice of the interaction potentials.

Initially, we study the (1+1)-dimensional pinning model with Gaussian potentials. This model will then be extended to general interaction potentials being subject to some weak conditions. It turns out that in presence of the gradient interaction there is always a trivial phase transition, i.e. the critical pinning value is zero. This is remarkable, since in absence of the gradient interaction, it is known that the critical pinning value is strictly positive. Furthermore, we investigate the behavior of the pinning model in the proximity of an impenetrable membrane (wetting model). Here the polymer fluctuates above the reference plane that imposes a repulsive effect. We show that in this case a proper phase transition occurs. Moreover, we treat (1+d)-dimensional models with Gaussian potentials. Here one has an additional choice between different pinning subspaces that play a crucial role for the localization behavior. Finally, we are interested in the phase transition and its order at the critical pinning value. In this context the regularity of the so called free energy is of great importance.

Contents

0   About this work   **11**
    0.1   Motivation 11
    0.2   Localization in terms of the free energy 13
    0.3   Existence of the free energy in our models 16
    0.4   Some notations 17

Outline   **19**

1   Gaussian pinning model   **21**
    1.1   Introduction and description of the model 21
    1.2   Free model in dependence of self-interaction parameters 22
    1.3   Related models and the main result 26
    1.4   Construction of a Markov chain 27
        1.4.1   The construction 27
        1.4.2   Connection to the free model $\mathbb{P}_{0,N}$ 30
        1.4.3   Identification of the density of (W_{n-1}, W_n) 33
    1.5   Pinning and interaction 36
        1.5.1   The contact process 36
        1.5.2   Markov renewal description 37
    1.6   Accurate determination of F_s 38
        1.6.1   Hilbert-Schmidt property 38
        1.6.2   Zerner's theorem and proof of Proposition 1.14 39
    1.7   Identification of the free energy and proof of Thm. 1.1 42
        1.7.1   The double-contact process 42
        1.7.2   Proof of Theorem 1.1 43
    1.8   Alternative proof of trivial phase transition 45
    1.9   Modification of the pinning model 48

2   General pinning model   **49**
    2.1   Introduction, model, results 49
        2.1.1   Introduction 49
        2.1.2   The model 49
        2.1.3   The result 50
    2.2   Markovian description of the free model 51
    2.3   Bounds on the density $\varphi_N^{(0,0)}(0,0)$ 56
        2.3.1   The lower bound 56
        2.3.2   The upper bound 58
    2.4   The "double-contact" process 59

2.5   Lower bound for the free energy 60

3   The polymer above a "hard wall" : entropic repulsion in Gaussian case 63
3.1   The model with a wall 63
3.2   The main result on wetting 64
3.3   The free wetting model 65
    3.3.1   Convenient representation for the integrated Markov chain 66
3.4   Entropic repulsion 67
    3.4.1   Upper bound 69
    3.4.2   Lower bound 77
3.5   Impact of pinning in wetting-model 79
    3.5.1   The contact process 79
    3.5.2   Construction of a semi Markov (sub)-kernel 80
3.6   Accurate determination of F_s^w 81
    3.6.1   An useful bound and the Hilbert-Schmidt property 82
    3.6.2   Perron-Frobenius Thm. and proof of Proposition 3.19 86
3.7   Identification of the free energy and proof of Thm. 3.1 87
    3.7.1   Obtaining an ordinary renewal process 87
    3.7.2   Proof of Theorem 3.1 88

4   Phase Transitions for higher dimensional Gaussian models 91
4.1   The model in higher dimensions 91
4.2   Free partition function 92
4.3   First example: heavy pinning 93
4.4   The weak pinning 94
    4.4.1   Compactness for wetting and high-dimensional pinning 96
    4.4.2   Compactness for pinning in lower dimensions 97
    4.4.3   Conclusion and results 103
    4.4.4   Remark on the Laplacian model 104
4.5   General pinning subspace M 105

5   Order of Phase Transitions for Gaussian models 107
5.1   Preliminaries and results 107
5.2   First order phase transition 109
5.3   Smooth phase transition 116
    5.3.1   The case of "proper" phase transition 116
    5.3.2   Proof of Theorem 5.2 in case of "proper" phase transition 122
    5.3.3   Theorem 5.2 and trivial phase transition 123

A   Appendix 129
A.1   Zerner's theorem 129
A.2   Toeplitz Matrices and determinants 129
A.3   Inverse elements of M_n 131
A.4   Recursive representation of $\{Y_i\}_{i\in\mathbb{Z}}$ 133
A.5   The conditional processes 135
A.6   The assumption 5.14 138

A.7 Some facts 140

References **144**

0 About this work

0.1 Motivation

We start our motivation from the point of view of polymers. This are chain-like molecules build up from small molecular units (monomers), which are connected by strong covalent bonds. There are several examples of matter consisting of polymers like: plastic, rubber or soap and some more complicated biopolymers like: cellulose, DNA or filaments, which form the cytoskeleton of cells. The variety and numerous applications of these objects interested originally chemists, biologists, physicists and material scientists. A situation of particular interest is the bundling of two (or more) polymers. For instance the stability of cytoskeletons of cells heavily depends on the bundling and number of interacting polymers. The second example is the so called DNA denaturation, which can be used to analyze some aspects of DNA. Here the two strands of the DNA are nothing else than polymers consisting of nucleotides as monomers. Another situation is the adsorption of a polymer on a planar substrate, for instance a polymer attracted to a membrane.

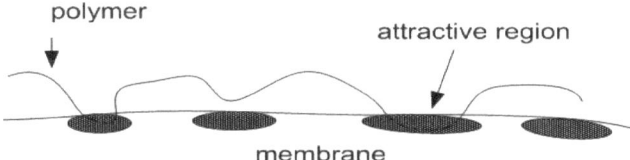

Figure 0.1: Polymer fluctuating nearby an impenetrable membrane with attractive regions.

Recently also mathematicians showed an active interest in a stochastic description of polymers. In order to develop such a probabilistic model for a polymer (random polymer) on an abstract level we have to extract some crucial properties. The following certainly belong to it

- stochastic spatial distribution
- self-avoidance
- interaction within the polymer
- interaction with environment .

Whereas those properties should be clear, it is worth to make a comment on the second point. The self-avoidance refers to the property of excluded-volume, which means that one part of a long chain molecule can not occupy space that is already occupied by another part of the same molecule. This seems to be trivial from physical point of view, however mathematically it causes a lot of trouble. The reason is that such a random polymer can be naturally described by self-avoiding walks. However, this models are eminently complicated to approach, especially when an interaction with the environment has to be taken into account. Instead one considers usually the so called directed walks. Here an additional deterministic component for the direction is introduced to obtain artificially the property of self avoidance. From the mathematical point of view the directed walks enable a much more farreaching analysis and application of technics and lead to interesting and challenging questions.

In this work we consider a (1+d)-dimensional direct model for a polymer chain, given by $i \to \varphi_i \in \mathbb{R}^d$, i.e. the position of the i-th monomer. The polymer is subject to an interaction with itself as well as an interaction with the environment. The environment has an attractive effect and is represented by an m-dimensional subspace (reference-plane). By touching the reference-plane the polymer receives a reward $\varepsilon \geq 0$. The interaction among individual monomers is given by the Hamilton operator with zero boundary conditions

$$\mathcal{H}_{[-1,N+1]}(\varphi_{-1},...,\varphi_{N+1}) = \frac{\alpha}{2}\sum_{i=1}^{N+1} V_1(\nabla \varphi_i) + \frac{\beta}{2}\sum_{i=0}^{N} V_2(\Delta\varphi_i) \ .$$

We call this model the pinning model and study particularly the localization behavior of the polymer in proximity of a reference plane. In other words, we are asking if the polymer sticks close to the reference plane (localization) or fluctuates away from it (delocalization). This can be seen as a phase transition, which occurs by modifying the force of attraction in the parameter ε. In this case we would have a critical value ε_c, such that for $\varepsilon > \varepsilon_c$ the polymer chain has a positive contact fraction at the reference plane, i.e. localization. On the other hand it would not be the case for $\varepsilon < \varepsilon_c$, i.e. delocalization takes place. This behavior addionally depends on the choice of the parameters α, β and the interaction potentials V_1, V_2. At the criticality the behavior has to be investigated separately and is closely connected to some regularity properties, which depend on the model. Furthermore we study also the so called wetting model. This corresponds to the situation in figure 0.1, where addionally a hard wall (membrane) is present and the chain is not allowed to enter into it.

From the physical point of view the models described above appear also in context of semiflexible polymers, cf. [19]. Here the bending rigidity of the polymer can be expressed in terms of the persistence length L_p. It is known that on length scales $L \leq L_p$ there occurs a semiflexible behavior, i.e. a rigid straight shape dominates. On the other hand on length scales $L \gg L_p$ large polymers appear to be flexible. Another example appears for instance in [24], here it serves as a model for interacting surfaces. The parameter α denotes the lateral tension and β the bending rigidity of some membranes. For just two interacting membranes there is a rigidity-dominated regime for sufficiently small scales and a tension-dominated regime for sufficiently large scales. In the last case the bending potential V_2 becomes irrelevant.

In mathematical papers the model for a flexible polymer ($\beta = 0$) is well known and has been studied by various authors, cf. [1],[7], [12], [13], [16], [18], [20]. The case of the pure Laplacian model ($\alpha = 0$) has been studied by Deuschel and Caravenna in [10] and [11] as a model for semiflexible polymers. In [11] also results on scaling limits have been obtained.

The purpose of the thesis is to combine both cases and study a new model with a mixed gradient and Laplacian interaction. The emphasis lies on localization and delocalization phenomena and we present new results in this context, cf. outline of the thesis. So far I am not aware of authors who studied the particular model in this thesis in any way. Nevertheless, [10] is the most important paper for the methods developed in this thesis. Coming back to the physical example above, a semiflexible polymer which is much longer than its persistence length L_p behaves effectively as a flexible chain of loosely connected rigids segments of size L_p, cf. also [9]. This behavior can be recovered in our simulations in chapter 1. However, observe that in our analysis we will take $N \to \infty$ and therefore the flexible behavior should prevail. This is indeed the case, as will be shown later.

0.2 Localization in terms of the free energy

In order to capture the phenomenon of localization and delocalization described above, one is inherently interested in a quantity that corresponds to each situation. Indeed, a fluctuating polymer in the proximity of an attractive region (and not only) can be described by the so called free energy F. This quantity should therefore reflect the behavior of "long" polymers in dependence of all relevant parameters of any kind of involved interaction. How could one define such an F ? Since we are dealing with an object that is motivated from statistical mechanics, it is known that the so called partition function plays an important role in this context. The partition function is just a normalizing quantity for the Gibbs-measures and so in discrete lattice-models it represents all the possible configurations of the corresponding model. Therefore it fulfills the demand on the free energy to capture the whole system and one can expect a relationship of both quantities.

Let us be more precise and relate the situation described above to the specific models that we study in this thesis. Throughout the thesis the spatial distribution of polymer-chains $\varphi^{(N)} := \{\varphi_1, ..., \varphi_{N-1}\} \in \mathbb{R}^{N-1}$ of finite length $N - 1$ is basically given by measures of type

$$\mathbb{P}_{\varepsilon,N}(d\varphi^{(N)}) := \frac{\exp(-\mathcal{H}_N(\varphi^{(N)}))}{\mathcal{Z}_{\varepsilon,N}} \prod_{i=1}^{N-1} (\varepsilon\delta_0(d\varphi_i) + d\varphi_i) \ . \qquad (0.1)$$

We remark at this point that this model will be modified in various ways in the following chapters. There we will explain the corresponding model in detail. Nevertheless it is worth to consider first (0.1) to getting started and relate some important properties. To explain the model briefly we denote by $d\varphi_i$ the Lebesgue measure on $\mathbb{R}$ and by $\delta_0(.)$ the Dirac mass at zero. Moreover the Hamiltonian $\mathcal{H}_N(.)$, which will be specified later, describes the self-interaction of the chain. This interaction determines the shape of the free model ($\varepsilon = 0$), i.e. when no external impact is present. The parameter $\varepsilon \geq 0$ is called pinning parameter and reflects a force that tries to pull down the chain towards the x-axis, which is also called the defect-line. In that sense we are dealing with a directed model that fulfills the properties of a polymer being attracted to some regions in the environment, confer section 0.1.

We return to the importance of the partition function, that we have already mentioned in the beginning of this section. In our case it is the normalizing quantity

$$\mathcal{Z}_{\varepsilon,N} := \int_{\mathbb{R}^{N-1}} \frac{\exp(-\mathcal{H}_N(\varphi^{(N)}))}{\mathcal{Z}_{\varepsilon,N}} \prod_{i=1}^{N-1} (\varepsilon \delta_0(d\varphi_i) + d\varphi_i) \ .$$

To capture the competing behavior of the fluctuations of the free model and the interaction with the environment, which is represented by pinning to the defect-line, we introduce finally

Definition 0.1 (Free energy)

$$F(\varepsilon) := \lim_{N \to \infty} F_N(\varepsilon) \quad and \quad F_N(\varepsilon) := \frac{1}{N} \log \left(\frac{\mathcal{Z}_{\varepsilon,N}}{\mathcal{Z}_{0,N}} \right) \quad , \ \varepsilon \geq 0 \ .$$

The free energy F is well defined by a super-additivity argument, cf. (0.3) for existence in our models. In the following we will give some properties of the free energy and explain why this quantity is related to localization. For this purposes we define the number of contacts to the defect line as

$$\ell_N := \#\{k \in \{1, ..., N\} \, | \, \varphi_k = 0\} \ .$$

Observe that under $\mathbb{P}_{\varepsilon,N}$ it holds $\ell_N \geq 1$, since $\varphi_N = 0$. By Lemma 5.3 and its proof we know that in our case the distribution of ℓ_N can be written as follows

$$\mathbb{P}_{\varepsilon,N}(\ell_N = k+1) = \frac{\varepsilon^k}{\mathcal{Z}_{\varepsilon,N}} R_{k,N} \quad \text{and so} \quad \mathcal{Z}_{\varepsilon,N} = \sum_{k=0}^{N-1} \varepsilon^k R_{k,N} \ . \tag{0.2}$$

Setting now $\widetilde{F}_N(t) := F_N(e^t)$ and $\widetilde{F}(t) := F(e^t)$, $t \in \mathbb{R}$, it is a simple computation to obtain

$$\widetilde{F}'_N(t) = \frac{1}{N} \mathbb{E}_{\mathbb{P}_{e^t,N}}[\ell_N - 1] \quad \text{and} \quad \widetilde{F}''_N(t) = \frac{1}{N} \text{Var}_{\mathbb{P}_{e^t,N}}[\ell_N - 1] \geq 0 \ .$$

Therefore $\widetilde{F}_N(.)$ is convex and so is the limit $\widetilde{F}(.)$. Hence $F(\varepsilon) = \widetilde{F}(\log \varepsilon)$ is continuous, as long as it is finite. We have scaled the free energy in such a way that at the origin the free energy of the free model ($\varepsilon = 0$) is zero. Let us remark an obvious, but important property

$$\mathcal{Z}_{\varepsilon,N} \geq \mathcal{Z}_{0,N} \quad \Longrightarrow \quad F(\varepsilon) \geq F(0) = 0 \ .$$

Since $\mathcal{Z}_{\varepsilon,N}$ is non-decreasing in ε, the same holds for F. So one could ask whether there exists an ε_c such that $F(\varepsilon)$ becomes strictly positive for all $\varepsilon > \varepsilon_c$. Indeed, we will see later on that this is true and hence the following definition is meaningful.

0.2 Localization in terms of the free energy

Definition 0.2 *The polymer measure $\mathbb{P}_{\varepsilon,N}$ is called*
- *delocalized, if $\varepsilon \in \mathcal{D} := \{\varepsilon \geq 0 \,|\, F(\varepsilon) = 0\}$ and*
- *localized, if $\varepsilon \in \mathcal{L} := \{\varepsilon \geq 0 \,|\, F(\varepsilon) > 0\}$.*

The point $\varepsilon_c := \sup\{\varepsilon \,|\, \varepsilon \in \mathcal{D}\}$ is called the critical point.

We will speak of a (proper) phase transition, when ε_c is strictly positive. Otherwise, if $\varepsilon_c = 0$, we call the phase transition trivial. However, due to the monotonicity and continuity of F at the moment it is only clear that $0 \leq \varepsilon_c \leq \infty$ and that localization can only occur if $\varepsilon > \varepsilon_c$. It seems that speaking of phase transition just by distinguish whether F is strictly positive or zero is quite high-toned. Nevertheless, we will see now that there is more behind it.

To this purpose we recall a fact from [10] that also fits in our setting. Namely, the relationship of F to the path-behavior of the polymer. We remark that in all our models F will be differentiable for every $\varepsilon \neq \varepsilon_c$, hence in this case we can set $d_\varepsilon := \varepsilon F'(\varepsilon) \geq 0$. Now for every $x \geq 0$ and $K > 0$ we can estimate by the Markov-inequality

$$\mathbb{P}_{\varepsilon,N}(\ell_N/N > d_\varepsilon + K) \leq e^{-x(d_\varepsilon+K)N} \, \mathbb{E}_{\mathbb{P}_{\varepsilon,N}}\left[e^{x\ell_N}\right] \, . \tag{0.3}$$

We will show that the r.h.s. decays exponentially in N. Observe now that by applying (0.2)

$$\frac{1}{N} \log \mathbb{E}_{\mathbb{P}_{\varepsilon,N}}\left[e^{x\ell_N}\right] = \frac{1}{N} \log \left(\sum_{k=0}^{N-1} e^{(k+1)x} \mathbb{P}_{\varepsilon,N}(\ell_N = k+1)\right)$$

$$= \frac{1}{N} \log \left(e^x \frac{1}{\mathcal{Z}_{\varepsilon,N}} \sum_{k=0}^{N-1} (e^x \varepsilon)^k R_{k,N}\right) = \frac{1}{N} \log \left(e^x \frac{\mathcal{Z}_{0,N}}{\mathcal{Z}_{\varepsilon,N}} \frac{\mathcal{Z}_{e^x\varepsilon,N}}{\mathcal{Z}_{0,N}}\right)$$

$$\xrightarrow[N\to\infty]{} F(e^x \varepsilon) - F(\varepsilon) \, .$$

Further on, by setting $h(x) := F(e^x \varepsilon)$ and performing Taylor-expansion in $x = 0$

$$h(x) = h(0) + h'(0)(x-0) + o(x^2) = F(\varepsilon) + (e^x \varepsilon F'(e^x\varepsilon))|_{x=0}(x-0) + o(x^2)$$
$$= F(\varepsilon) + \varepsilon x F'(\varepsilon) + o(x^2) = F(\varepsilon) + d_\varepsilon x + o(x^2)$$

and therefore

$$\lim_{N\to\infty} \frac{1}{N} \log \left(e^{-x(d_\varepsilon+K)N} \mathbb{E}_{\mathbb{P}_{\varepsilon,N}}\left[e^{x\ell_N}\right]\right) = -x(d_\varepsilon + K) + d_\varepsilon x + o(x^2) = -xK + o(x^2) \, .$$

Consequently for every $K > 0$ there exists a constant $c > 0$ such that for all $N \in \mathbb{N}$

$$\mathbb{P}_{\varepsilon,N}(\ell_N/N > d_\varepsilon + K) \leq e^{-cN} \, .$$

Analogously one can show $\mathbb{P}_{\varepsilon,N}(\ell_N/N < d_\varepsilon - K) \leq e^{-cN}$. Recall that this holds for each $\varepsilon \neq \varepsilon_c$, since we already mentioned that here d_ε is well defined. Consequently we can distinguish between two cases:

a) If $\varepsilon < \varepsilon_c$ then $d_\varepsilon = 0$ and for every $K > 0$ there exists an $c > 0$ such that

$$\mathbb{P}_{\varepsilon,N}(\ell_N/N > K) \leq e^{-cN} \quad , \quad \text{for all } N \in \mathbb{N} . \qquad (0.4)$$

b) If $\varepsilon > \varepsilon_c$ then $d_\varepsilon > 0$ and for every $K > 0$ there exists an $c > 0$ such that

$$\mathbb{P}_{\varepsilon,N}(|\ell_N/N - d_\varepsilon| > K) \leq e^{-cN} \quad , \quad \text{for all } N \in \mathbb{N} . \qquad (0.5)$$

Although at the critical point $F(\varepsilon_c) = 0$, it is not at all clear which of the cases above should be the right one here. We defer this question to chapter 5, where the differentiability at the critical point is of special interest.

Case a) and b) tell us in which way the paths behave on the basis of the contact fraction ℓ_N/N. In case a) the typical paths touch the x-axis just in a sub-linear way, i.e. $o(N)$-times. Whereas in case b) the typical paths touch the x-axis linearly with the contact fraction d_ε. Thus, except at criticality, the differentiation between $F(\varepsilon) = 0$ or $F(\varepsilon) > 0$ diplays the crucial difference also in this paths-sense and in this context the term of phase transition in ε_c should be justified.

As we have already mentioned, the model (0.1) will be modified in different directions. Nevertheless, exactly the same considerations as above can be made for all our models that we treat in the thesis. Also the definitions of localization/delocalization and the free energy transfer directly to the extended models choosing the appropriate partition function $\mathcal{Z}_{\varepsilon,N}$.

0.3 Existence of the free energy in our models

Although the existence of the free energy in our models will be implicitly ensured by its construction, we will give here the classical approach on how to prove that fact. It is usually provided through a supper-additivity argument. Let $\mathcal{Z}_{\varepsilon,N}$ be a representative for all partition functions of the mixed model that we consider in this thesis. First of all by expansion of the product measure (1.22) and restricting only to $A = \{N, N+1\}$ we obtain for all $N, M \geq 1$

$$\mathcal{Z}_{\varepsilon,N+M+1} \geq \varepsilon^2 \mathcal{Z}_{\varepsilon,N} \mathcal{Z}_{\varepsilon,M}$$

which, by fixing ε and setting $Z_N := \mathcal{Z}_{\varepsilon,N-1}$, is equivalent to $Z_{N'+M'} \geq \varepsilon^2 Z_{N'} Z_{M'}$. Here we have set $N' := N+1$ and $M' := M+1$. Therefore we have

$$\log(Z_{n+m}) \geq 2\log(\varepsilon) + \log(Z_n) + \log(Z_m) . \qquad (0.6)$$

Setting $Z'_n := \log(Z_n) + 2\log(\varepsilon)$ by a well known Lemma of Fakete the limit of Z'_n/n ($n \to \infty$) exists (possibly infinite), since by (0.6) we have the super-additivity condition $Z'_{n+m} \geq Z'_n + Z'_m$. Hence of course also

$$\lim_{N \to \infty} \frac{1}{N} \log \mathcal{Z}_{\varepsilon,N}$$

exists for all $\varepsilon > 0$. The limit for $\varepsilon = 0$ will be investigated separately in later chapters.

0.4 Some notations

In the following we give some notations that are frequently used throughout the thesis.

In case of existence, let us define for sequences a_n and b_n the notation

$$a_n = o(b_n) :\Longleftrightarrow \lim_{n\to\infty} \frac{a_n}{b_n} = 0 \, ,$$

$$a_n = O(b_n) :\Longleftrightarrow \lim_{n\to\infty} \frac{a_n}{b_n} = c \, , \ |c| \neq 0 \ (\text{ and } |c| \neq \infty)$$

$$a_n \sim b_n :\Longleftrightarrow \lim_{n\to\infty} \frac{a_n}{b_n} = 1 \, ,$$

and furthermore for two positive sequences $\{c_n\}_{n\in\mathbb{N}}$ and $\{d_n\}_{n\in\mathbb{N}}$

$$c_n \succeq d_n \quad :\Longleftrightarrow \quad \lim_{n\to\infty} \frac{c_n}{d_n} \geq c \quad , \text{ for some } c > 0$$

and

$$c_n \preceq d_n \quad :\Longleftrightarrow \quad \lim_{n\to\infty} \frac{c_n}{d_n} \leq \tilde{c} \quad , \text{ for some } \tilde{c} \geq 0 \, .$$

The number-sets we use are $\mathbb{Z}^+ := \{0, 1, 2, ...\}$, $\mathbb{N} := \{1, 2, ...\}$ and $\mathbb{R}^+ := [0, \infty)$. For $t \in \mathbb{R}$ the lower (upper) integer part $\lfloor t \rfloor$ ($\lceil t \rceil$) is the largest (smallest) integer smaller (larger) or equal than t. Sometimes we use in calculations constants in a wide sense, meaning that a constant represents just a constant expression, independently of its value. This happens usually in long calculations and by that we simply avoid an introduction of numerous constants.

Let us denote as usual by $\mathcal{B}(\mathbb{R})$ the Borel σ-field of $\mathbb{R}$. We call a function $K_{.,.} : \mathbb{R} \times \mathcal{B}(\mathbb{R}) \to \mathbb{R}^+$ a σ-finite kernel, if

- for every $x \in \mathbb{R}$ the $K_{x,.}$ is a σ-finite Borel-measure on $\mathbb{R}$ and
- for every $A \in \mathcal{B}(\mathbb{R})$ the $K_{.,A}$ is a Borel-measurable function.

Let K, G be two σ-finite kernels, then we define

- the composition by $(K \circ G)_{x,dy} := \int_{z \in \mathbb{R}} K_{x,dz} G_{z,dy}$,
- the n-fold self composition by $K^{\circ n}_{x,dy}$,
- the 0-fold composition by $K^{\circ 0}_{x,dy} := \delta_x(dy)$ and
- the sum of compositions $(1-K)^{-1}_{x,dy} := \sum_{n=0}^{\infty} K^{\circ n}_{x,dy}$.

We consider also σ-finite kernels that additionally depend on $n \in \mathbb{Z}^+$, i.e. $K_{x,dy}(n)$. Analogously for two kernels $K_{.,.}(.), G_{.,.}(.)$ of this type we define

- the convolution by

$$(K * G)_{x,dy}(n) := \sum_{i=0}^{n} (K(i) \circ G(n-i))_{x,dy} = \sum_{i=0}^{n} \int_{z \in \mathbb{R}} K_{x,dz}(i) G_{z,dy}(n-i) \, ,$$

- the m-fold self convolution of $K_{x,dy}(n)$ by $K_{x,dy}^{*m}(n)$ and
- the 0-fold convolution by $K_{x,dy}^{*0} := \delta_x(dy)(n)\mathbb{1}_{\{n=0\}}$.

Let c_n be a positive sequence and $\asymp \in \{\sim, \preceq, \succeq\}$. In order to capture the asymptotical behavior between kernels $K_{.,.}(.), G_{.,.}$ we denote by

$$K_{x,dy}(n) \asymp \frac{G_{x,dy}}{c_n} :\Longleftrightarrow K_{x,B}(n) \asymp \frac{G_{x,B}}{c_n} \quad , \text{ as } n \to \infty$$

for every $x \in \mathbb{R}$ and every bounded set $B \in \mathcal{B}(\mathbb{R})$.

Outline

This work is structured in the following way:

In chapter 1 the $(1+1)$-dimensional pinning model is studied as a first model with gradient and Laplacian mixture type interaction and Gaussian potentials. In the beginning we consider some simulations for the free model in dependence of α and β. The parameters tune the strength of influence that each potential contributes. Then we investigate the connection of the free model to some integrated Markov chain and treat in detail the representation of the density for the two last steps of the chain. We then prove a trivial phase transition ($\varepsilon_c = 0$) for this model. The approach via Markov renewal theory is based on ideas developed by Deuschel and Caravenna for the Laplacian case in [10]. We also extend the idea of [7] to present an alternative way to obtain a trivial phase transition. Finally we make a comment on the critical absence time from the defect-line for a model with modified pinning strength.

Chapter 2 is devoted to studying the pinning model with general interaction potentials. It turns out that even for strong potentials of the Laplacian interaction the transitions stays still trivial. This is remarkable in view of results in [10], where for the pure Laplacian model a (non-trivial) phase transition has been proven. This result has been obtained during my stay in Padova supervised by Francesco Caravenna. The proof is based on a lower bound for the free energy and requires sufficient estimates on the inter-arrival law.

The Gaussian wetting model is treated in chapter 3. Here we introduce the additional effect of a wall, which the chain is not allowed to cross. This effect is of repulsive character and we can indeed prove that there is a phase transition with strictly positive critical point. The approach here is similar to that in chapter 1, however we give an explicit representation for the Markov chain, which simplifies some arguments. This representation is mainly used to handle the so called entropic repulsion that is crucial for the investigation of this model. Here we use a "decoupling-argument" that allows us to extract a random walk from the integrated Markov chain. This argument is based on Gaussian properties and is the reason why we cannot extend the results to non-Gaussian potentials in that way.

In chapter 4 we consider higher dimensional Gaussian pinning and wetting models. Here we have the additional choice in the dimension of the pinning subspace, which alters the localization behavior substantially. We treat heavy and weak pinning spaces (referring to the possible extremal dimension of the pinning subspace) and then a general pinning subspace, which can be seen as a combination of both. Our analysis yields the same results as in the pure gradient case [7]. The approach is similar as in chapter 1 and our main concern will be compactness criteria of some integral operators. A comment on the $(1 + d)$-dimensional Laplacian model will be also given in this context.

Finally in chapter 5 we study the regularity of the phase transition in the critical regime. It turns out that higher dimensional models display a discontinuity, if considering the fraction of times the Polymer touches the defect-subspace. For the weak-pinning models we have proven a first order phase transition for pinning in $d \geq 5$ and wetting in $d \geq 3$. Moreover the transition is of second order for the pinning model in $d = 3$ and $d = 4$. Under some assumptions we have treated also the remaining lower dimensions. Similar to [10], the investigation of the first moment of the double-contact process in the model at criticality is the key in the proof. At the end, in the Appendix we give some calculations and technichal results needed in this book.

1 Gaussian pinning model

1.1 Introduction and description of the model

We consider a (1+1)-dimensional model, i.e. a directed model for a linear chain, which is described by its configurations $\{(n, \varphi_n)\}_{0 \leq n \leq N}$. The chain is randomly distributed in space and undergoes an interaction with the environment and itself. Thus, it can be seen as a so called random polymer and we want to study its spatial distribution as a function of its length and its interaction parameters. The selfinteraction consists of a Gradient and Laplacian mixture type. Whereas the interaction with the environment will be reduced to a δ-pinning, i.e. the chain gets a reward $\varepsilon \geq 0$ by touching the x-axis (defect-line). We are going to discuss the localization behavior, which was already introduced and motivated in chapter 0. As we will see this behavior is substantially different, depending on the parameters α, β and $\varepsilon \geq 0$.

We are going to explain the model in more detail now. For $j, k \in \mathbb{Z}$ with $k - j \geq 2$ and $\varphi \in \mathbb{R}^{k-j+1}$ consider the Hamiltonian

$$\mathcal{H}_{[j,k]}(\varphi) := \mathcal{H}_{[j,k]}^{(1)}(\varphi) + \mathcal{H}_{[j,k]}^{(2)}(\varphi)$$

where

$$\mathcal{H}_{[j,k]}^{(1)}(\varphi) = \sum_{i=j+2}^{k} V_1(\nabla \varphi_i)$$

$$\mathcal{H}_{[j,k]}^{(2)}(\varphi) = \sum_{i=j+1}^{k-1} V_2(\Delta \varphi_i) .$$

The interaction potentials $V_1(\eta) = \alpha V(\eta)$ and $V_2(\eta) = \beta V(\eta)$ are defined by $V : \mathbb{R} \to \mathbb{R}$, $\eta \mapsto \eta^2/2$, for some later in Remark 1.4 specified admissible constants α, β that are fixed. Furthermore ∇ and Δ denote the discrete gradient

$$\nabla \varphi_n := \varphi_n - \varphi_{n-1}$$

and discrete Laplace operator

$$\Delta \varphi_n := \nabla \varphi_{n+1} - \nabla \varphi_n = \varphi_{n+1} + \varphi_{n-1} - 2\varphi_n .$$

In this chapter we are interested in the following pinning model, which is given by the spatial distribution on $\mathbb{R}^{N-1}$:

$$\mathbb{P}_{\varepsilon,N}(d\varphi) := \frac{\exp(-\mathcal{H}_{[-1,N+1]}(\varphi))}{\mathcal{Z}_{\varepsilon,N}} \prod_{i=1}^{N-1} (\varepsilon\delta_0(d\varphi_i) + d\varphi_i) \;, \qquad (1.1)$$

where the Hamiltonian can be written now in the form

$$\mathcal{H}_{[-1,N+1]}(\varphi_{-1}, ..., \varphi_{N+1}) = \frac{\alpha}{2} \sum_{i=1}^{N+1} (\nabla\varphi_i)^2 + \frac{\beta}{2} \sum_{i=0}^{N} (\Delta\varphi_i)^2$$

and for simplicity we impose zero boundary conditions, i.e.

$$\varphi_{-1} = \varphi_0 = \varphi_N = \varphi_{N+1} = 0 \;.$$

Furthermore, the remaining quantities in the pinning model (1.1) denote

- $\varepsilon \geq 0$ the pinning parameter
- $\delta_0(.)$ the Dirac mass at zero
- $d\varphi_i$ the Lebesgue measure on $\mathbb{R}$
- $\mathcal{Z}_{0,N}$ the normalization constant (partition function).

The interaction with the environment is reduced to a δ-pinning at the x-axis and so the chain is rewarded by touching this defect-line. We remark that this model undergoes two opposite effects, the entropy and the energy, represented by the self-interaction and the δ-pinning. Both effects can be strengthened or weakened by varying the parameters α, β and ε.

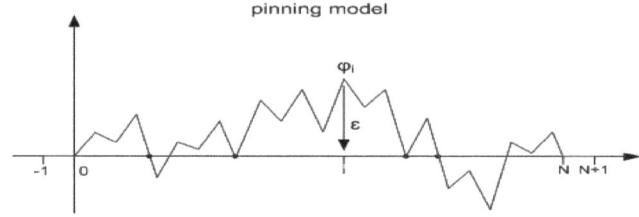

Figure 1.1: This is a sketch of the pinning model $\mathbb{P}_{\varepsilon,N}$. The polymer is represented by the heights φ_i of monomers, which are attracted to the interger-sites at the x-axis. The black points denote the contacts to those sites. Of course, here it is possible for the polymer to overcome the defect line without getting any reward.

1.2 Free model in dependence of self-interaction parameters

As a first step it is worth to gain a feeling on how the free model ($\varepsilon = 0$) behaves by choosing different self-interaction parameters α and β. Let us first take a look at some

1.2 FREE MODEL IN DEPENDENCE OF SELF-INTERACTION PARAMETERS

simulations. For the sake of convenience, except in figure 1.2, we have chosen free boundary conditions on the right.

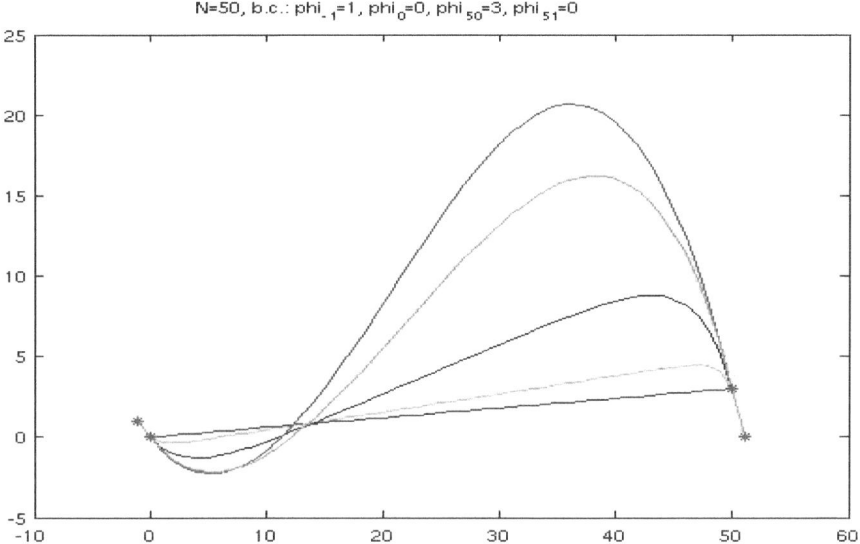

Figure 1.2: We have simulated here five minimizers (most favored paths) for the model (1.1) with the boundary conditions $\varphi_{-1} = 1, \varphi_0 = 0, \varphi_N = 3, \varphi_{N+1} = 0$ and $N = 50$. It is clear that for zero boundary conditions all minimizers would be $\equiv 0$. The colors correspond to $(\alpha = 1, \beta = 0), (\alpha = 1, \beta = 1), (\alpha = 1, \beta = 10), (\alpha = 1, \beta = 100)$ and $(\alpha = 0, \beta = 1)$.

Here we can see the crucial difference between the gradient case ($\beta = 0$) and the Laplacian one ($\alpha = 0$). The gradient model (blue) favors a direct path without carrying about smoothness. In contrast, the Laplacian model (red) is rigid to bendings and prefers a smooth path. In between we have fixed α and by increasing β the path of the Laplacian model can be approximated. Therefore one can speak (like in physical literature) of flexible polymers in gradient and of semi-flexible polymers in the Laplacian case.

Gaussian pinning model

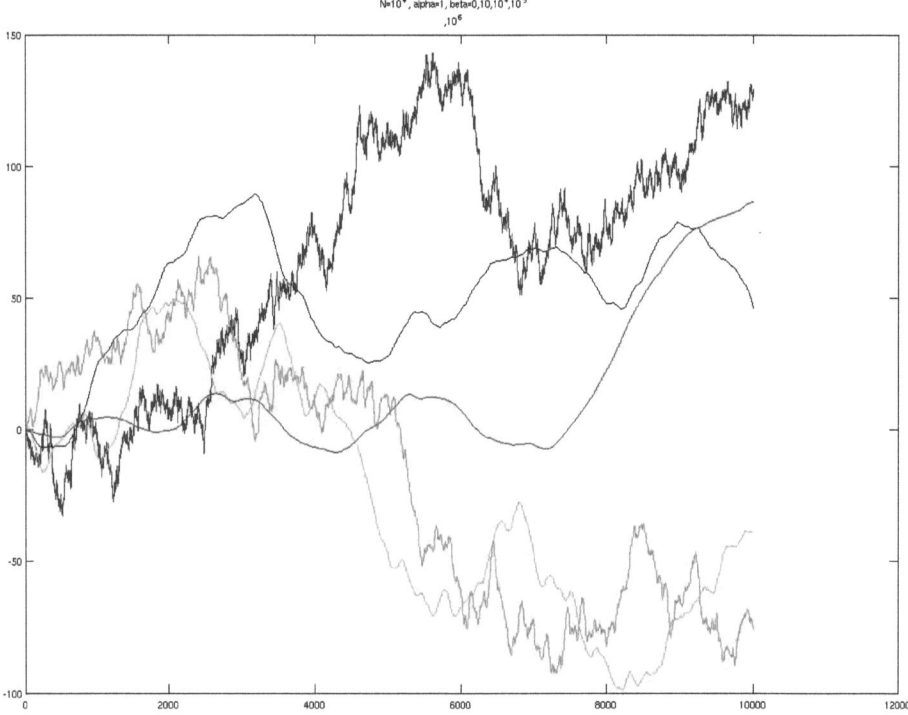

Figure 1.3: This is a simulation of some trajectories of $\mathbb{P}_{0,N}$ with free boundary conditions on the right and $N = 10^4$. The colors correspond to $(\alpha = 1, \beta = 0), (\alpha = 1, \beta = 100), (\alpha = 1, \beta = 10^4), (\alpha = 1, \beta = 10^5)$ and $(\alpha = 1, \beta = 10^6)$.

The blue trajectory, corresponding to the gradient model, is the most jagged one. In fact, we will see later that it is nothing else than a random walk trajectory. Now, by increasing β the trajectories become smoother, cf. the red one with $\beta = 10^6$. We didn't simulate a trajectory of the Laplacian model in the same figure, since this takes place on different height-scales, as can be seen by the next picture.

1.2 FREE MODEL IN DEPENDENCE OF SELF-INTERACTION PARAMETERS

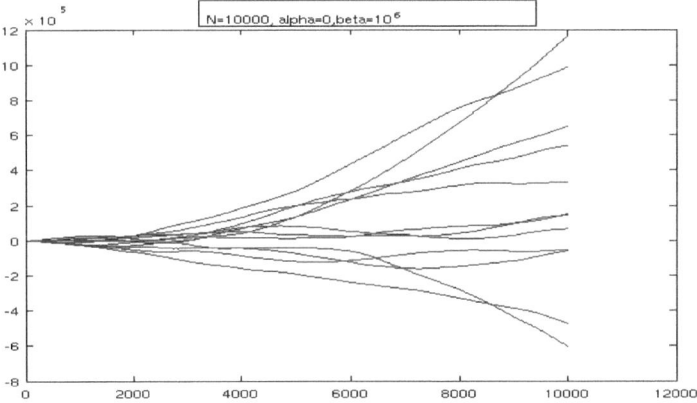

Figure 1.4: This is a simulation of twelve trajectories of $\mathbb{P}_{0,N}$ with free boundary conditions on the right and $N = 10^4$. The parameters are $\alpha = 0, \beta = 10^6$.

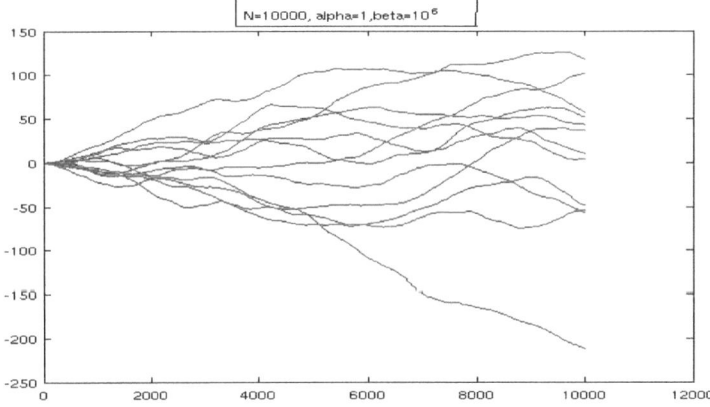

Figure 1.5: This is a simulation of twelve trajectories of $\mathbb{P}_{0,N}$ with free boundary conditions on the right and $N = 10^4$. The parameters are $\alpha = 1, \beta = 10^6$.

At a first glance the comparison of figure 1.4 and figure 1.5 seems to suggest a similar smoothness behavior. However, observe that by the height-range one could conjecture different behavior on variances of φ_N. In figure 1.4 one can read off $\text{Var}(\varphi_N) \approx N^3$ and in figure 1.5 $\text{Var}(\varphi_N) \approx N$. These is indeed true, as will be seen later on.

1.3 Related models and the main result

Some natural questions arise in the context of the competing behavior between entropy (fluctuations of the free model) and the energy (interaction with the defect-line):

Q1: Is the energy large enough to pin the chain at the x-axis ?

Q2: Does a critical point exist, where below entropy and above energy prevail ?

It is known that the model behaves dramatically different for the two extremal cases $\alpha = 0$ and $\beta = 0$. For instance, the gradient case has been studied by [7] for Gaussian potentials, cf. also [1], [12], [13], [16], [18], [20]. Whereas the Laplacian case was investigated for general potentials in [10], which is the most important paper for the methods developed in this thesis. In the gradient case the phase transition was proven to be trivial and for the Laplacian model it is proper, meaning $\varepsilon_c > 0$. Consequently one is immediately inclined to ask whether there are some critical values for α and β where the model changes its behavior. Recall the meaning of localization/delocalization in chapter 0 and the free energy

$$F(\varepsilon) = \lim_{N \to \infty} \frac{1}{N} \log \left(\frac{\mathcal{Z}_{\varepsilon,N}}{\mathcal{Z}_{0,N}} \right) \quad , \; \varepsilon \geq 0 \,. \tag{1.2}$$

We were able to prove the following localization result for the model (1.1)

Theorem 1.1 (Localization for the pinning case) *For every $\alpha, \beta > 0$ the model $\mathbb{P}_{\varepsilon,N}$ exhibits a trivial phase transition $\varepsilon_c = 0$, i.e.*

$$\mathcal{D} = \{0\} \quad \text{and} \quad \mathcal{L} = (0, \infty) \,.$$

Furthermore, on $\mathcal{L}$ the free energy is real analytic and

$$F(\varepsilon) \sim \log \varepsilon \quad , \; \varepsilon \to \infty \,.$$

In terms of localization the model reflects the behavior of the gradient model, as was mentioned above. It is remarkable that even a huge $\beta \gg 0$ in the Laplacian part has no influence on change in the localization behavior. This shows a very strong impact of the ∇-interaction in our model (1.1). Therefore in the limit $N \nearrow \infty$ our model behaves effectively as a flexible chain, although on small scales it is rigid to bendings. Thus, it can be called also semi-flexible, cf. section 0.1.

1.4 Construction of a Markov chain

1.4.1 The construction

First of all we would like to describe the simplified situation where no pinning is present, i.e. the free model $\mathbb{P}_{0,N}$. We will see that there is a connection to a specific Markov chain, which will be constructed in what follows.

For $x \in \mathbb{R}$ and $f \in L^2(\mathbb{R}, dx)$ we define the operator

$$K(x, f) := Kf(x) := \int f(y)\, k(x,y)\, dy \qquad (1.3)$$

$$\text{and} \quad k(x,y) = e^{-V_1(y)-V_2(y-x)}\ .$$

It is an compact operator on $L^2(\mathbb{R}, dx)$. Indeed, it is even a Hilbert-Schmidt operator, due to the quadratic potentials V_1, V_2 and assumption **(AP)** in Remark 1.4 below

$$\int_\mathbb{R} \int_\mathbb{R} k(x,y)^2\, dx\, dy = \int_\mathbb{R} e^{-2V_1(y)}\, dy \int_\mathbb{R} e^{-2V_2(x)}\, dx < \infty\ .$$

Thanks to the infinite dimensional Perron-Frobenius Theorem A.1, there exists an almost surely strictly positive right eigenfunction $\nu \in L^2(\mathbb{R}, dx)$ to the largest eigenvalue $\lambda > 0$. This fact enables us to construct a certain process, which will be very useful later on. More precisely the construction goes as follows.

For $a, b \in \mathbb{R}$ we consider a probability space $(\Omega, \mathcal{A}, \mathrm{P}^{(a,b)})$ and two processes $\{Y_i\}_{i \in \mathbb{Z}^+}$, $\{W_i\}_{i \in \mathbb{Z}^+}$ with the properties:

- $\{Y_i\}_{i \in \mathbb{Z}^+}$ is a Markov process with $Y_0 = a$ and the transition probability

$$\mathrm{P}^{(a,b)}(Y_{n+1} = dy | Y_n = x) \sim k(x,y) \frac{\nu(y)}{\lambda\, \nu(x)}\, dy \qquad (1.4)$$

  and

- $\{W_i\}_{i \in \mathbb{Z}^+}$ is the integrated Markov process with

$$W_0 = b \quad \text{and} \quad W_n = b + Y_1 + ... + Y_n.$$

Of course, the transition probability has to be well defined and therefore the a.s. positiveness of ν is not enough. Nevertheless rewriting

$$\nu(x) = \frac{1}{\lambda} \int_\mathbb{R} \nu(y)\, k(x,y)\, dy$$

one sees that also $\nu(x) > 0$ for all $x \in \mathbb{R}$. Since we are in a Gaussian setting much more can be said on the quantities defined above, cf. Proposition 1.5.

Remark 1.2
The model (1.1) is well defined when $\alpha \geq 0$ and additionally $\alpha + 4\beta \geq 0$ (but of course not $\alpha = \beta = 0$). This can be seen by writing the Hamiltonian like in the proof of proposition 1.11

$$\mathcal{H}_{[-1,n+1]}(0,0,w^T,0,0) = \frac{1}{2}\langle w, (\alpha A_{n-1} + \beta B_{n-1})w\rangle .$$

If $\alpha, \beta \geq 0$ (but not $\alpha = \beta = 0$), then $\alpha A_n + \beta B_n$ is positive definite for all $n \in \mathbb{N}$, because A_n and B_n have this property. Let us consider now the case $\alpha > 0, \beta < 0$ and $\alpha + 4\beta > 0$. Here we have

$$\mathcal{H}_{[-1,n+1]}(0,0,w^T,0,0) = \mathcal{H}^{(1)}_{[-1,n+1]}(0,0,w^T,0,0) + \frac{\beta}{2}\sum_{i=0}^{n}(w_{i+1}+w_{i-1}-2w_i)^2$$

$$\geq \mathcal{H}^{(1)}_{[-1,n+1]}(0,0,w^T,0,0) + \frac{\beta}{2}\sum_{i=0}^{n}\left[2(w_{i+1}-w_i)^2 + 2(w_i - w_{i-1})^2\right]$$

$$= \frac{1}{2}\langle w, (\alpha + 4\beta)A_{n-1}w\rangle$$

and so in this case $\alpha A_n + \beta B_n$ is positive definite for all $n \in \mathbb{N}$. In the last case $\alpha > 0, \beta < 0$ and $\alpha + 4\beta = 0$ we have computed the determinant, cf. appendix A.2

$$\det(\alpha A_n + \beta B_n) = \frac{1}{8}\left[\beta^n + (-\beta)^n(7 + 8n + 2n^2)\right]$$

which is also positive for all $n \in \mathbb{N}$ and so implies the positive definiteness. For the sake of completeness, for $\alpha = 0$ and $\beta = 0$ we have

$$\det(\beta B_n) = \frac{1}{12}\beta^n(2+n)^2(3+4n+n^2) \quad \text{and} \quad \det(\alpha A_n) = \alpha^n(n+1) .$$

Remark 1.3
At a first glance, considering the inequalities between the Hamiltonians above, one is inclined to compare the gradient, Laplacian and mixed model just over the corresponding partition functions to obtain easily a statement about the free energy. However this is a deception regarding definition (1.2) of the free energy as a ratio.

Remark 1.4
Observe that for the extremal values $\alpha = 0$, $\beta = 0$ or $\alpha + 4\beta = 0$ the statements we make later on are not always defined, but you can get the according results by passing to the limit $\alpha \to 0$, $\beta \to 0$ or $\alpha \to -4\beta$. Nevertheless, from now on we will assume that

(AP) $\boxed{\alpha, \beta > 0}$.

This is not a restriction on parameters and it prevents some unnecessary case differentiations, cf. chapter 2 and Remark A.6.

As already mentioned above, in the Gaussian case we can obtain a much more detailed information about the transition probability:

1.4 Construction of a Markov chain

Proposition 1.5 *The spectral radius of K and its corresponding eigenfunction ν have the following explicit representation*

$$\lambda = \left(\frac{2\pi}{\sigma_+}\right)^{1/2} \quad , \quad \sigma_+ = \frac{\alpha + 2\beta + \sqrt{\alpha}\sqrt{\alpha + 4\beta}}{2}$$

and

$$\nu(x) = e^{x^2(\alpha - \sqrt{\alpha}\sqrt{\alpha + 4\beta})/4} \quad , \quad x \in \mathbb{R} .$$

Before proving this we will need a small

Lemma 1.6 *If any eigenfunction $r \in L^2(\mathbb{R}, dx)$ of K has the property $r(x) > 0$ a.s. then its corresponding eigenvalue has to be the spectral radius of K.*

Proof This Lemma can be of course extended to any compact operator on $L^p(\mathbb{R}, dx)$, $1 \leq p \leq \infty$, taking r in the appropriate space.
Now wlog consider any right-eigenfunction $r \in L^2(\mathbb{R}, dx)$ with the property in the Lemma and the corresponding eigenvalue M, i.e.

$$\int_\mathbb{R} k(x,y)\, r(y)\, dy = M\, r(x) .$$

Furthermore consider the corresponding left eigenfunction $l \in L^2(\mathbb{R}, dx)$ to the spectral radius λ of K, i.e.

$$\int_\mathbb{R} k(x,y)\, l(x)\, dx = \lambda\, l(y) .$$

By Zerner's Theorem A.1 we know that $l(x) > 0$ a.s., therefore it follows

$$\lambda \int_\mathbb{R} l(y) r(y)\, dy = \int_\mathbb{R} \left(\int_\mathbb{R} k(x,y)\, l(x)\, dx\right) r(y)\, dy = \int_\mathbb{R} l(x)\, M\, r(x)\, dx .$$

Due to $\int l(y) r(y)\, dy > 0$ this means that $\lambda = M$. $\square$

This Lemma enables us now to approach

Proof of Propsition 1.5

We will first calculate the explicit form of an eigenvalue $\tilde{\lambda}$ of K and its corresponding eigenfunction $\tilde{\nu}$ and afterwards prove that they equal λ and ν. We make a guess on $\tilde{\nu}$ and take as an ansatz a quadratic function $\tilde{\nu}(x) = \exp\{ry^2 + sy + t\}$ with r, s, t to be specified. Now for $2r < \alpha + \beta$ one can calculate

$$K(x, e^{ry^2 + sy + t}) = \frac{\sqrt{2\pi}}{\sqrt{\alpha + \beta - 2r}} \exp\left\{\frac{x^2(2\beta r - \alpha\beta) + 2\beta s x + s^2 + 2\alpha t + 2\beta t - 4rt}{2(\alpha + \beta - 2r)}\right\}. \tag{1.5}$$

To verify the eigenvalue equation $K\tilde{\nu}(x) = \tilde{\lambda}\tilde{\nu}(x)$ we compare both sides and obtain:

$$r_{1,2} = \frac{\alpha \pm \sqrt{\alpha^2 + 4\alpha\beta}}{4} \quad , \quad s = 0 \quad , \quad t \in \mathbb{R} .$$

We choose $t = 0$ to simplify the multiplicative constant of the eigenvalue and $r := r_2$ to have (recall $\beta > 0$) $\tilde{\nu} \in L^2(\mathbb{R}, dx)$. Consequently, from (1.5) an eigenvalue of K and its corresponding eigenfunction are

$$\tilde{\lambda} = \frac{\sqrt{2\pi}}{\sqrt{\alpha + \beta - 2r_2}} = \left(\frac{2\pi}{\sigma_+}\right)^{1/2} \quad \text{and} \quad \tilde{\nu}(x) = e^{r_2 x^2} = e^{x^2(\alpha - \sqrt{\alpha}\sqrt{\alpha + 4\beta})/4} .$$

Clearly $\tilde{\nu}(x) > 0$ for all $x \in \mathbb{R}$ and therefore Lemma 1.6 tells us that $\tilde{\lambda}$ indeed equals λ, the spectral radius of K and so it holds $\nu = \tilde{\nu}$. □

Remark 1.7
Similarly to the last proof one can compute the left eigenfunction $w \in L^2(\mathbb{R}, dx)$ of K to the spectral radius λ

$$w(x) = \frac{\sqrt{\sqrt{\alpha}\sqrt{\alpha + 4\beta}}}{\sqrt{2\pi}} e^{x^2(-\alpha - \sqrt{\alpha}\sqrt{\alpha + 4\beta})/4} ,$$

i.e. $\int k(x,y) w(x) dx = \lambda w(y)$, $y \in \mathbb{R}$. It holds $\langle \nu, w \rangle_{L^2(\mathbb{R}, dx)} = 1$. The invariant distribution of the Markov chain $\{Y_n\}_n$ is

$$\pi(dx) := \nu(x) w(x) dx = \frac{\sqrt{\sqrt{\alpha}\sqrt{\alpha + 4\beta}}}{\sqrt{2\pi}} e^{-x^2(\sqrt{\alpha}\sqrt{\alpha + 4\beta})/2} dx , \qquad (1.6)$$

as one can convince hisself by direct computation.

1.4.2 Connection to the free model $\mathbb{P}_{0,N}$

Next, let us take a look at the finite dimensional distribution of our Markov chain $\{W_i\}_{i \in \mathbb{Z}^+}$.

Proposition 1.8 *For $n \in \mathbb{N}$ and $w_{-1} := b - a$, $w_0 := b$ we have*

$$\mathrm{P}^{(a,b)}\left((W_1, ..., W_n) \in (dw_1, ..., dw_n)\right) = \frac{\nu(w_n - w_{n-1})}{\lambda^n \nu(a)} e^{-\mathcal{H}_{[-1,n]}(w_{-1},...,w_n)} \prod_{i=1}^{n} dw_i . \qquad (1.7)$$

Proof Under $\mathrm{P}^{(a,b)}$ we have already set $Y_n = W_n - W_{n-1}$, $n \geq 1$, so the law of $(W_1, ..., W_n)$ is determined by the law of $(Y_1, ..., Y_n)$. If we set $y_i := w_i - w_{i-1}$, $i \geq 2$ and $y_1 := w_1 - b$, then we have to show that under the r.h.s. of (1.7) the $(y_i)_{i=1,...,n}$ are distributed like the first n steps of a Markov chain starting at a with the transition probability given by (1.4). The Hamiltonian can be now written in the following way

$$\mathcal{H}_{[-1,n]}(w_{-1}, ..., w_n) = \sum_{i=1}^{n} V_1(y_i) + V_2(y_1 - a) + \sum_{i=1}^{n-1} V_2(y_{i+1} - y_i) .$$

Therefore we conclude

$$\frac{\nu(w_n - w_{n-1})}{\lambda^n \nu(a)} e^{-\mathcal{H}_{[-1,n]}(w_{-1},...,w_n)} = \frac{\nu(y_n)}{\lambda^n \nu(a)} k(a, y_1) \prod_{i=2}^{n} k(y_{i-1}, y_i)$$

$$= \frac{\nu(y_1)}{\lambda \nu(a)} k(a, y_1) \prod_{i=2}^{n} \frac{\nu(y_i)}{\lambda \nu(y_{i-1})} k(y_{i-1}, y_i) ,$$

1.4 CONSTRUCTION OF A MARKOV CHAIN

and we are done, because the last statement is just the density of the law of $(Y_1, ..., Y_n)$ under $P^{(a,b)}$ w.r.t. the Lebesgue-measure $dy_1 \cdots dy_n$. □

Observe that like in the Laplacian-model by the last proposition under $P^{(a,b)}$ the integrated Markov chain $\{W_i\}_{i \in \mathbb{Z}^+}$ is a process with memory two. Whereas the combined process $\{(Y_i, W_i)\}_{i \in \mathbb{Z}^+}$ is a Markov process starting in $(Y_0, W_0) = (a, b)$. The next quantity will play an important role in the further analysis.

Definition 1.9 *For $n \geq 2$ we define the density of (W_{n-1}, W_n) by*

$$\varphi_n^{(a,b)}(w_1, w_2) := \frac{P^{(a,b)}((W_{n-1}, W_n) \in (dw_1, dw_2))}{dw_1 dw_2}. \tag{1.8}$$

The following statement connects the free model $\mathbb{P}_{0,N}$ to our constructed Markov chain.

Proposition 1.10

$$\mathbb{P}_{0,N}(.) = P^{(0,0)}((W_1, ..., W_{N-1}) \in . | W_N = W_{N+1} = 0) \tag{1.9}$$

and

$$\mathcal{Z}_{0,N} = \lambda^{N+1} \varphi_{N+1}^{(0,0)}(0, 0).$$

Proof With the help of (1.7), the r.h.s. of (1.9) can be written (conditional density)

$$P^{(0,0)}((W_1, ..., W_{N-1}) \in . | W_N = W_{N+1} = 0)$$
$$= \frac{1}{\lambda^{N+1} \varphi_{N+1}^{(0,0)}(0,0)} \cdot \int_{.} e^{-\mathcal{H}_{[-1,N+1]}(w_{-1}, ..., w_{N+1})} \prod_{i=1}^{N-1} dw_i,$$

where $w_{-1} = w_0 = w_N = w_{N+1} = 0$. The first expression in the above calculation is a probability measure, so plugging in $\mathbb{R}^{N-1}$ we have

$$\iff \int_{\mathbb{R}^{N-1}} e^{-\mathcal{H}_{[-1,N+1]}(w_{-1},...,w_n)} \prod_{i=1}^{N-1} dw_i = \lambda^{N+1} \varphi_{N+1}^{(0,0)}(0,0)$$
$$\iff \mathcal{Z}_{0,N} = \lambda^{N+1} \varphi_{N+1}^{(0,0)}(0,0).$$

This concludes the proof. □

Here we see that our free model is just the law of the integrated Markov chain conditioned on $\{W_N = W_{N+1} = 0\}$, i.e. a "bridge" of the process $\{W_i\}_{i \in \mathbb{Z}^+}$.

We have already seen the importance of the free energy as an indicator for the behavior of the model. We will first study the normalizing part of the free energy. More precisely, we are interested in the asymptotical behavior of the free partition function $\mathcal{Z}_{0,N}$.

Proposition 1.11 *We have the following limits for the free partition function*

$$\sqrt{n-1} \left(\frac{\sigma_+}{2\pi}\right)^{\frac{n-1}{2}} \mathcal{Z}_{0,n} \xrightarrow[n \to \infty]{} c, \quad c \in (0, \infty)$$

and therefore
$$\frac{1}{n}\log \mathcal{Z}_{0,n} \xrightarrow[n\to\infty]{} \log \lambda \qquad (1.10)$$
where σ_+ was defined in Proposition 1.5.

Proof Let $w = (w_1, ..., w_{n-1}^T)$ and $w_{-1} = w_0 = w_n = w_{n+1} = 0$, then it is easily seen that the Hamiltonian can be written as a quadratic form
$$\mathcal{H}_{[-1,n+1](0,0,w^T,0,0)} = \frac{1}{2}\langle w, (\alpha A_{n-1} + \beta B_{n-1})w\rangle$$
where $A_n, B_n \in \mathbb{R}^{n\times n}$ are

$$A_n = \begin{pmatrix} 2 & -1 & 0 & \cdots & 0 \\ -1 & \ddots & \ddots & \ddots & \vdots \\ 0 & \ddots & \ddots & \ddots & 0 \\ \vdots & \ddots & \ddots & \ddots & -1 \\ 0 & \cdots & 0 & -1 & 2 \end{pmatrix}, \quad B_n = \begin{pmatrix} 6 & -4 & 1 & 0 & \cdots & 0 \\ -4 & 6 & -4 & 1 & \ddots & \vdots \\ 1 & \ddots & \ddots & \ddots & \ddots & 0 \\ 0 & \ddots & \ddots & \ddots & \ddots & 1 \\ \vdots & \ddots & \ddots & \ddots & \ddots & -4 \\ 0 & \cdots & 0 & 1 & -4 & 6 \end{pmatrix}$$

This enables us to write
$$\mathcal{Z}_{0,n} = \int_{\mathbb{R}^{n-1}} e^{-\langle w,(\alpha A_{n-1}+\beta B_{n-1})w\rangle/2} \prod_{i=1}^{n-1} = \left(\frac{(2\pi)^{n-1}}{\det(\alpha A_{n-1}+\beta B_{n-1})}\right)^{1/2}.$$

Now we are left with the problem of finding a nice representation of the determinant of a matrix, which depends on the variables α, β and the size parameter n of the matrix. Finding such a representation has cost us a lot of time, but finally we discovered that for the so called finite Toeplitz matrices there are methods which give certain representation for the determinants, cf. [8]. We defer the calculation to appendix A.2 and this method gives
$$\det(\alpha A_{n-1} + \beta B_{n-1}) = c_1^{\alpha,\beta}\sigma_+^{n-1} + c_2^{\alpha,\beta}\sigma_+^{n-1}(n-1) + o(\sigma_+^{n-1}) \qquad (1.11)$$
with some constants which we computed exactly, but we indicate here only the crucial one
$$c_2^{\alpha,\beta} = \frac{2\beta^2\sqrt{\alpha} + \alpha^2\sqrt{\alpha+4\beta} + \alpha^{5/2} + 2\alpha\beta\sqrt{\alpha+4\beta} + 4\sqrt{\alpha}\alpha\beta}{2\alpha^2\sqrt{\alpha+4\beta}}.$$
So
$$\sqrt{n-1}\left(\frac{\sigma_+}{2\pi}\right)^{\frac{n-1}{2}}\mathcal{Z}_{0,n} = \left(\frac{n-1}{c_1^{\alpha,\beta}+c_2^{\alpha,\beta}(n-1)+o(1)}\right)^{1/2} \xrightarrow[n\to\infty]{} \left(\frac{1}{c_2^{\alpha,\beta}}\right)^{1/2} =: c$$
and therefore
$$\lim_{n\to\infty}\frac{1}{n}\log\mathcal{Z}_{0,n} = \lim_{n\to\infty}\frac{n-1}{2n}\log\left(\frac{2\pi}{\sigma_+}\right) = \frac{1}{2}\log\left(\frac{2\pi}{\sigma_+}\right).$$

$\square$

1.4.3 Identification of the density of (W_{n-1}, W_n)

We turn to the density $\varphi_n^{(a,b)}$ defined in (1.8). It appeared already in connection with the free partition function in Proposition 1.10 and will play further on an important role when interaction with the defect line is present. The next Proposition gives us an explicit form of this density.

Proposition 1.12 *We have the following three statements:*

(i) *The density defined in* (1.8) *is Gaussian:* $\varphi_n^{(a,b)} \sim \mathcal{N}(m_n^{\alpha,\beta}(a,b), \Sigma_n)$. *The expectation is* $\mathbb{R}^2 \ni m_n^{\alpha,\beta}(a,b) := (\mu_{n-1}^{\alpha,\beta}(a,b), \mu_n^{\alpha,\beta}(a,b))^T$ *where*

$$\mu_{n-1}^{\alpha,\beta}(a,b) = \frac{\sqrt{\alpha}(2b-a) + a\sqrt{\alpha+4\beta} + a(\frac{\sigma_-}{\beta})^n(-\sqrt{\alpha}-\sqrt{\alpha+4\beta})}{2\sqrt{\alpha}} \quad , \quad (1.12)$$

$$\mu_n^{\alpha,\beta}(a,b) = \frac{\sqrt{\alpha}(2b-a) + a\sqrt{\alpha+4\beta} + a(\frac{\sigma_-}{\beta})^n(\sqrt{\alpha}-\sqrt{\alpha+4\beta})}{2\sqrt{\alpha}} \quad (1.13)$$

and

$$\sigma_- := \frac{\alpha + 2\beta - \sqrt{\alpha}\sqrt{\alpha+4\beta}}{2} \quad , \quad -1 < \frac{\sigma_-}{\beta} < 1 \; . \quad (1.14)$$

The (2×2)-*covariance matrix has the form* $\Sigma_n = R_n M_n^{-1} R_n^T$, *where*

$$R_n = \begin{pmatrix} 0 & \cdots & 0 & 1 & 0 \\ 0 & \cdots & 0 & 0 & 1 \end{pmatrix} \in \mathbb{R}^{2 \times n}$$

and $\mathbb{R}^{n \times n} \ni M_n = \alpha \tilde{A}_n + \beta \tilde{B}_n + (\sqrt{\alpha}\sqrt{\alpha+4\beta}/2 - \alpha/2)C_n$ *with*

$$\tilde{A}_n = \begin{pmatrix} 2 & -1 & 0 & \cdots & & 0 \\ -1 & \ddots & \ddots & \ddots & \ddots & \vdots \\ 0 & \ddots & \ddots & \ddots & \ddots & \vdots \\ \vdots & \ddots & \ddots & \ddots & \ddots & 0 \\ 0 & \cdots & 0 & -1 & 2 & -1 \\ 0 & \cdots & \cdots & 0 & 1 & 1 \end{pmatrix} \quad , \quad C_n = \begin{pmatrix} 0 & \cdots & & & \cdots & 0 \\ \vdots & \ddots & \ddots & & \ddots & \vdots \\ \vdots & & \ddots & 0 & 0 & 0 \\ \vdots & & & \ddots & 0 & 1 & -1 \\ 0 & \cdots & & & 0 & -1 & 1 \end{pmatrix}$$

and

$$\tilde{B}_n = \begin{pmatrix} 6 & -4 & 1 & 0 & 0 & \cdots & & 0 \\ -4 & 6 & -4 & 1 & 0 & \cdots & & 0 \\ 1 & \ddots & \ddots & \ddots & \ddots & \ddots & & \vdots \\ 0 & \ddots & \ddots & \ddots & \ddots & \ddots & \ddots & \vdots \\ \vdots & \ddots & \ddots & \ddots & \ddots & \ddots & \ddots & 0 \\ 0 & \cdots & 0 & 1 & -4 & 6 & -4 & 1 \\ 0 & \cdots & & 0 & 1 & -4 & 5 & -2 \\ 0 & \cdots & & \cdots & 0 & 1 & -2 & 1 \end{pmatrix} .$$

(ii) *The following asymptotics holds:* $\det \Sigma_n = O(n)$.

(iii) In particular $\varphi_n^{(0,0)}$ is a $\mathcal{N}(0, \Sigma_n)$ Gaussian density.

Proof We first look at the proof of (iii).
Let us take the left b.c. $w_{-1} = w_0 = 0$, the right w_{n-1}, w_n free and $w = (w_{-1}, ..., w_n)^T$, then

$$\nu(w_n - w_{n-1})e^{-\mathcal{H}_{[-1,n]}(w)} = \exp\left(-\frac{1}{2}\left[(w_n - w_{n-1})^2 \left(\frac{\sqrt{\alpha}\sqrt{\alpha + 4\beta} - \alpha}{2}\right) + \langle w, (\alpha\tilde{A}_n + \beta\tilde{B}_n)w\rangle\right]\right).$$

Therefore proposition 1.8 allows us to write

$$\bar{\varphi}_n^{(0,0)}(w_1, ..., w_n) := \frac{\mathrm{P}^{(0,0)}\left((W_1, ..., W_n) \in (dw_1, ..., dw_n)\right)}{dw_1 \cdots dw_n} = \frac{1}{\lambda^n} e^{-\langle w, M_n w\rangle/2}.$$

From proposition 1.8 we know that $\bar{\varphi}_n^{(0,0)}$ is a density, furthermore M_n is symmetric, so M_n has to be positive definite and $\bar{\varphi}_n^{(0,0)}$ is Gaussian. Now $(W_{n-1}, W_n) \sim \mathcal{N}(0, R_n M_n^{-1} R_n^T)$, because $(W_1, ..., W_n) \sim \mathcal{N}(0, M_n^{-1})$. Thus we can write

$$\varphi_n^{(0,0)}(w_{n-1}, w_n) = \frac{1}{2\pi\sqrt{\det(\Sigma_n)}} \exp\left\{-\frac{1}{2}\langle \begin{pmatrix} w_{n-1} \\ w_n \end{pmatrix}, \Sigma_n^{-1} \begin{pmatrix} w_{n-1} \\ w_n \end{pmatrix}\rangle\right\}$$

and by proposition 1.10 and proposition 1.11 we get

$$\frac{1}{2\pi\sqrt{\det(\Sigma_n)}} = \frac{\mathcal{Z}_{0,n-1}}{\lambda^n} = \frac{\sigma_+^{n/2}}{2\pi}\left(\frac{1}{\det(\alpha A_{n-2} + \beta B_{n-2})}\right)^{1/2} = \frac{\sigma_+}{2\pi}\left(\frac{1}{c_1^{\alpha,\beta} + c_2^{\alpha,\beta}(n-2) + o(1)}\right)^{1/2}.$$
(1.15)

But this implies (ii).
So it remains to show (i). For this purpose we take the left b.c. $w_{-1} = b-a$, $w_0 = b$, the right w_{n-1}, w_n free and $w = (w_{-1}, ..., w_n)^T$, then

$$\nu(w_n - w_{n-1})e^{-\mathcal{H}_{[-1,n]}(w)} = \exp\left(-\frac{1}{2}\left[(w_n - w_{n-1})^2\left(\frac{\sqrt{\alpha}\sqrt{\alpha+4\beta}-\alpha}{2}\right) + 2\mathcal{H}_{[-1,n]}(w)\right]\right).$$
(1.16)

Next we will try to obtain a quadratic form in (1.16), so we denote by $H_{[-1,n]}^{(a,b)}(w)$ the expression $[\cdots]$. Due to symmetric matrices it can be written in the way

$$H_{[-1,n]}^{(a,b)}(w) = \alpha\langle w - \mu_\nabla, \tilde{A}_n(w-\mu_\nabla)\rangle + \beta\langle w-\mu_\Delta, \tilde{B}_n(w-\mu_\Delta)\rangle$$
$$+ (w_n - w_{n-1})^2\left(\frac{\sqrt{\alpha}\sqrt{\alpha+4\beta}-\alpha}{2}\right)$$
$$= \langle w, M_n w\rangle - 2\langle w, \alpha\tilde{A}_n\mu_\nabla + \beta\tilde{B}_n\mu_\Delta\rangle + \alpha\langle \mu_\nabla, \tilde{A}_n\mu_\nabla\rangle + \beta\langle \mu_\Delta, \tilde{B}_n\mu_\Delta\rangle,$$
(1.17)

where $\mu_\nabla, \mu_\Delta \in \mathbb{R}^n$ and

$$\mu_\nabla = (b, ..., b)^T \quad, \quad \mu_\Delta = (b+a, b+2a, ..., b+na)^T.$$

Our aim is to find μ, Υ (of course dependent on n, a, b, α and β), s.th.

$$H_{[-1,n]}^{(a,b)}(w) \stackrel{!}{=} \langle w - \mu, M_n(w-\mu)\rangle + \Upsilon.$$
(1.18)

1.4 Construction of a Markov chain

Comparing (1.17) and (1.18) and using the symmetry of M_n, we have

$$-2\langle w, M_n\mu\rangle \stackrel{!}{=} -2\langle w, \alpha\tilde{A}_n\mu_\nabla + \beta\tilde{B}_n\mu_\Delta\rangle\ ,$$

so $M_n\mu = \alpha\tilde{A}_n\mu_\nabla + \beta\tilde{B}_n\mu_\Delta$ and therefore

$$\mu = M_n^{-1}(\alpha\tilde{A}_n\mu_\nabla + \beta\tilde{B}_n\mu_\Delta)\ . \tag{1.19}$$

Furthermore this implies

$$\Upsilon = \alpha\langle\mu_\nabla, \tilde{A}_n\mu_\nabla\rangle + \beta\langle\mu_\Delta, \tilde{B}_n\mu_\Delta\rangle - \langle\mu, M_n\mu\rangle\ . \tag{1.20}$$

Observe that to compute μ_{n-1}, μ_n and Υ, we need only 8 elements of M_n^{-1}, because

$$\tilde{A}_n\mu_\nabla = (b, 0, ..., 0)^T\quad ,\quad \tilde{B}_n\mu_\Delta = (3b+a, -b, 0, ..., 0)^T\ .$$

It took us quite a while, but in the end we found a paper by Rózsa, who describes in [26] how to compute the inverse of some banded matrices. We defer the calculation to appendix A.3. This method was quite costly, but finally we were able to compute exactly the means in (1.12),(1.13) and

$$\Upsilon = a^2\left(\frac{\sqrt{\alpha}\sqrt{\alpha+4\beta}-\alpha}{2}\right)\ .$$

Now, from (1.16), (1.18) and proposition 1.8 it follows by $e^{-\Upsilon/2}/\nu(a) = 1$ that

$$\frac{\mathrm{P}^{(a,b)}((W_1, ..., W_n) \in (dw_1, ..., dw_n))}{dw_1 \cdots dw_n} = \frac{1}{\lambda^n}e^{-\langle w-\mu, M_n(w-\mu)\rangle/2}\ .$$

Finally we conclude that $(W_{n-1}, W_n) \sim \mathcal{N}(R_n\mu, R_nM_n^{-1}R_n^T)$ under $\mathrm{P}^{(a,b)}$, so

$$\varphi_n^{(a,b)}(w_{n-1}, w_n) = \frac{1}{2\pi\sqrt{\det(\Sigma_n)}}\exp\left\{-\frac{1}{2}\left\langle\begin{pmatrix}w_{n-1}-\mu_{n-1}^{\alpha,\beta}(a,b)\\w_n-\mu_n^{\alpha,\beta}(a,b)\end{pmatrix}, \Sigma_n^{-1}\begin{pmatrix}w_{n-1}-\mu_{n-1}^{\alpha,\beta}(a,b)\\w_n-\mu_n^{\alpha,\beta}(a,b)\end{pmatrix}\right\rangle\right\}\ . \tag{1.21}$$

The last thing is to show (1.14). If $\alpha, \beta > 0$,

$$-1 = \frac{\alpha+2\beta-\sqrt{\alpha+4\beta}\sqrt{\alpha+4\beta}}{2\beta} < \frac{\sigma_-}{\beta} < \frac{\alpha+2\beta-\sqrt{\alpha}\sqrt{\alpha}}{2\beta} = 1\ .$$

If $\alpha > 0, \beta < 0$ and $\alpha + 4\beta \geq 0$, then

$$\frac{\sigma_-}{\beta} = \frac{\sqrt{\alpha}\sqrt{\alpha+4\beta}-(\alpha+2\beta)}{2|\beta|}$$

and so

$$-1 = \frac{\sqrt{\alpha+4\beta}\sqrt{\alpha+4\beta}-(\alpha+2\beta)}{2|\beta|} < \frac{\sigma_-}{\beta} < \frac{\sqrt{\alpha}\sqrt{\alpha}-(\alpha+2\beta)}{2|\beta|} = 1\ .$$

□

Remark 1.13 *From proposition 1.12 we know, that for all $a, b, x, y \in \mathbb{R}$ there exists $c_1 > 0$ such that*

$$\varphi_n^{(a,b)}(x,y) = \varphi_n^{(0,0)}(x - \mu_{n-1}^{\alpha,\beta}(a,b), y - \mu_n^{\alpha,\beta}(a,b)) \leq \frac{1}{2\pi\sqrt{\det\Sigma_n}} \leq \frac{c_1}{\sqrt{n}}\ .$$

1.5 Pinning and interaction

Up to now we have studied the free pinning model $\mathbb{P}_{0,N}$. Now it is time to approach a description where interaction comes into play and a "strength" attracts the chain at the x-axis. This describes exactly the model $\mathbb{P}_{\varepsilon,N}$ for an $\varepsilon > 0$.

1.5.1 The contact process

We define the contact process $(\tau_i)_{i \in \mathbb{Z}^+}$ by

$$\tau_0 := 0 \quad \text{and} \quad \tau_{i+1} := \inf\{k > \tau_i \,|\, \varphi_k = 0\}$$

and the process $(J_i)_{i \in \mathbb{Z}^+}$, which gives the height of the polymer before the contact points

$$J_0 := 0 \quad \text{and} \quad J_i := \varphi_{\tau_i - 1}\,.$$

Next, one can expand the product measure in the definition of $\mathbb{P}_{\varepsilon,N}$ in the following way. Let $M := \{1, ..., N-1\}$, then

$$\prod_{i=1}^{N-1} (\varepsilon \delta_0(d\varphi_i) + d\varphi_i) = \sum_{A \subseteq M} \varepsilon^{|A|} \left(\prod_{i \in A} \delta_0(d\varphi_i)\right) \left(\prod_{j \in M \setminus A} d\varphi_j\right)\,. \tag{1.22}$$

Set the number of contacts to the defect line as $\ell_N = \#\{i \in \{1,...,N\} \,|\, \varphi_i = 0\}$ or equivalently $\ell_N = \max\{k \,|\, \tau_k \leq N\}$. Now take for fixed $k \in \mathbb{N}$ a time-partition $(t_i)_{i=1,...,k} \in \mathbb{N}$ with $0 < t_1 < \cdots < t_{k-1} < t_k := N$. Moreover suppose the contacts are only at $\tau_i = t_i$, $i = 1,...,k-1$ and set in (1.22) $A = \{\tau_1,...,\tau_{k-1}\}$. In order to obtain now the joint law of the process $\{\ell_N, (\tau_i)_{i \leq \ell_N}, (J_i)_{i \leq \ell_N}\}$, one has to integrate over φ_i where $i \notin A \cup (A-1)$. More precisely, for $(y_i)_{i=1,...,k} \in \mathbb{R}$ it is

$$\mathbb{P}_{\varepsilon,N}(\ell_N = k, \tau_i = t_i, J_i \in dy_i, i = 1,...,k)$$
$$= \frac{\varepsilon^{k-1}}{\mathcal{Z}_{\varepsilon,N}} F_{0,dy_1}(t_1) F_{y_1,dy_2}(t_2 - t_1) \cdots F_{y_{k-1},dy_k}(N - t_{k-1}) F_{y_k,\{0\}}(1)\,, \tag{1.23}$$

where $F_{x,dy}(n) := f_{x,y}(n) \mu(dy)$, $\mu(dy) := \delta_0(dy) + dy$ and

$$f_{x,y}(n) := \begin{cases} e^{-\beta x^2/2} \mathbb{1}_{\{y=0\}} & , n = 1 \\ e^{-\mathcal{H}_{[-1,2]}(x,0,y,0)} \mathbb{1}_{\{y \neq 0\}} & , n = 2 \\ \int_{\mathbb{R}^{n-2}} e^{-\mathcal{H}_{[-1,n]}(w_{-1},...,w_n)} \mathbb{1}_{\{y \neq 0\}} dw_1 \cdots dw_{n-2} & , n \geq 3 \\ \text{with } w_{-1} = x, w_0 = 0, w_{n-1} = y, w_n = 0\,. \end{cases} \tag{1.24}$$

Next we set

$$\widetilde{f}_{x,y}(n) := \frac{\nu(-y)}{\lambda^n \nu(-x)} f_{x,y}(n) \quad \text{and} \quad \widetilde{F}_{x,dy}(n) := \widetilde{f}_{x,y}(n) \mu(dy)\,.$$

Because of (1.7) we have

$$\varphi_n^{(a,b)}(x,y) = \frac{\nu(y-x)}{\lambda^n \nu(a)} \int_{\mathbb{R}^{n-2}} e^{-\mathcal{H}_{[-1,n]}(w_{-1},...,w_n)} \prod_{i=1}^{n-2} dw_i$$

where $w_{-1} = b - a, w_0 = b, w_{n-1} = x$ and $w_n = y$. Therefore for $x, y \in \mathbb{R}$ and $n \geq 2$

$$\widetilde{f}_{x,y}(n) = \varphi_n^{(-x,0)}(y,0)\mathbb{1}_{\{y \neq 0\}} . \tag{1.25}$$

1.5.2 Markov renewal description

We are going to describe the process of contact points in a more exact way. Our model consists of three-body interaction terms, therefore it won't be possible to describe $(\tau_i)_{i \in \mathbb{Z}^+}$ by a renewal process. Nevertheless something else can be proven, but first we define

$$K_{x,dy}^\varepsilon(n) := \varepsilon \widetilde{F}_{x,dy}(n) e^{-F_s(\varepsilon) n} \frac{\nu_\varepsilon(y)}{\nu_\varepsilon(x)} , \tag{1.26}$$

for an F_s and ν, which are specified by the next proposition.

Proposition 1.14 *For every $\varepsilon > 0$ there exist $F_s(\varepsilon) \in (0, \infty)$ and $\nu(\varepsilon) \in (0, \infty)$ with the property*

$$\int_{y \in \mathbb{R}} \sum_{n \in \mathbb{N}} K_{x,dy}^\varepsilon(n) = 1 \quad, \text{ for all } x \in \mathbb{R} . \tag{1.27}$$

We postpone the proof to section 1.6, where a lot more can be said about F_s and ν and even an explicit representation can be given.
This proposition is very useful, because it says that $K_{\cdot,\cdot}^\varepsilon(\cdot)$ denotes just a semi Markov kernel. The probabilistic interpretation of such kernels is the fact that one can now define a law $\mathbb{P}_\varepsilon$ under which $\{(\tau_i, J_i)\}_{i \in \mathbb{Z}^+}$ is a Markov chain on $\mathbb{Z}^+ \times \mathbb{R}$ with $(\tau_0, J_0) = (0, 0)$ and the transition kernel

$$\mathbb{P}_\varepsilon((\tau_{i+1}, J_{i+1}) \in (\{n\}, dy) \mid (\tau_i, J_i) = (m, x)) = K_{x,dy}^\varepsilon(n-m) . \tag{1.28}$$

Then the contact process $(\tau_i)_{i \in \mathbb{Z}^+}$ is called a Markov renewal process and $(J_i)_{i \in \mathbb{Z}^+}$ its modulating chain. The reason for this name comes from the fact that the increments $\{\tau_k - \tau_{k-1}\}_{k \in \mathbb{N}}$ are independent conditionally on $(J_i)_{i \in \mathbb{Z}^+}$, cf. [10]. Furthermore we can rewrite (1.23) as follows

$$\mathbb{P}_{\varepsilon,N}(\ell_N = k, \tau_i = t_i, J_i \in dy_i, i = 1,...,k)$$
$$= \frac{e^{F_s(\varepsilon)(N+1)}}{\varepsilon^2 \mathcal{Z}_{\varepsilon,N}} \lambda^{N+1} K_{0,dy_1}^\varepsilon(t_1) K_{y_1,dy_2}^\varepsilon(t_2 - t_1) \cdots K_{y_{k-1},dy_k}^\varepsilon(N - t_{k-1}) K_{y_k,\{0\}}^\varepsilon(1) \tag{1.29}$$

and for $t_0 = y_0 = 0$ the normalizing constant of $\mathbb{P}_{\varepsilon,N}$ has then to be

$$\mathcal{Z}_{\varepsilon,N} = \frac{e^{F_s(\varepsilon)(N+1)}}{\varepsilon^2} \lambda^{N+1} \sum_{k=1}^{N} \sum_{\substack{t_i \in \mathbb{N}, i=1,...,k \\ 0 < t_1 < \cdots < t_k := N}} \int_{\mathbb{R}^k} \left(\prod_{i=1}^{k} K_{y_{i-1},dy_i}^\varepsilon(t_i - t_{i-1}) \right) K_{y_k,\{0\}}^\varepsilon(1) .$$
$$\tag{1.30}$$

The next result reveals a connection between $\mathbb{P}_{\varepsilon,N}$, which is dependent on N, and $\mathcal{P}_\varepsilon$, which is not.

Proposition 1.15 *Define* $\mathcal{A}_N := \{\exists j \geq 0 \,|\, \tau_j = N, \tau_{j+1} = N+1\}$. *Then for all* $N \in \mathbb{N}, \varepsilon > 0$ *and* $k \leq N$ $((t_i)_{i=1,\ldots,k}, (y_i)_{i=1,\ldots,k}$ *as usual)*

$$\mathbb{P}_{\varepsilon,N}(\ell_N = k, \tau_i = t_i, J_i \in dy_i, \ i \leq k) = \mathcal{P}_\varepsilon(\ell_N = k, \tau_i = t_i, J_i \in dy_i, \ i \leq k \,|\, \mathcal{A}_N)$$

and

$$\mathcal{Z}_{\varepsilon,N} = \frac{e^{F_s(\varepsilon)(N+1)}}{\varepsilon^2} \lambda^{N+1} \mathcal{P}_\varepsilon(\mathcal{A}_N) \,. \tag{1.31}$$

Proof Due to (1.28) we have

$$\mathcal{P}_\varepsilon(\ell_N = k, \tau_i = t_i, J_i \in dy_i, \ i \leq k \,|\, \mathcal{A}_N)$$
$$= \frac{1}{\mathcal{P}_\varepsilon(\mathcal{A}_N)} K^\varepsilon_{0,dy_1}(t_1) K^\varepsilon_{y_1,dy_2}(t_2 - t_1) \cdots K^\varepsilon_{y_{k-1},dy_k}(N - t_{k-1}) K^\varepsilon_{y_k,\{0\}}(1)$$

and knowing that $\mathcal{P}_\varepsilon(. \,|\, \mathcal{A}_N)$ is a probability measure and comparing with (1.29) and (1.30) we arrive at the end of the proof. $\square$

We remark that this is an important observation, since we have obtained a connection between the contacts of the chain and a Markv renewal process conditioned on $\mathcal{A}_N$. Even more, all the dependence on N is "concentrated " just in the set $\mathcal{A}_N$.

1.6 Accurate determination of F_s

In this section we are going to prove Proposition 1.14 and give some explicit representations for the quantities therein. In particular, we show that F_s and ν_ε in (1.26) can be chosen such that (1.27) is fulfilled. The first step is the following Lemma, which is even more than we want.

1.6.1 Hilbert-Schmidt property

Lemma 1.16 *For every* $\theta > 0$ *the operator* $(B^\theta h)(x) := \int_\mathbb{R} B^\theta_{x,dy} h(y)$ *on the Hilbert-space* $L^2(\mathbb{R}, d\mu)$ *is a Hilbert-Schmidt operator, where* $B^\theta_{x,dy} := \sum_{n \in \mathbb{N}} e^{-\theta n} \widetilde{F}_{x,dy}(n)$.

Proof Let $\theta > 0$. We set $B^\theta_{x,dy} = b^\theta(x,y)\mu(dy)$ and

$$b^\theta(x,y) := e^{-\theta} \widetilde{f}_{x,0}(1) \mathbb{1}_{\{y=0\}} + \sum_{n \geq 2} e^{-\theta n} \widetilde{f}_{x,y}(n) \mathbb{1}_{\{y \neq 0\}} \,,$$

then we have to show

$$\int_\mathbb{R} \int_\mathbb{R} b^\theta(x,y)^2 \, \mu(dx)\, \mu(dy) \, < \infty \,.$$

It is

$$b^\theta(x,y)^2 = e^{-2\theta} \widetilde{f}_{x,0}(1)^2 \mathbb{1}_{\{y=0\}} + \sum_{n,m \geq 2} e^{-\theta(n+m)} \widetilde{f}_{x,y}(n) \widetilde{f}_{x,y}(m) \mathbb{1}_{\{y \neq 0\}}$$

and so

$$\int_{\mathbb{R}}\int_{\mathbb{R}} b^\theta(x,y)^2 \,\mu(dx)\,\mu(dy) = \int_{\mathbb{R}} e^{-2\theta}\widetilde{f}_{x,0}(1)^2 \,\mu(dx) + \sum_{n,m\geq 2} e^{-\theta(n+m)} \int_{\mathbb{R}} \widetilde{f}_{0,y}(n)\widetilde{f}_{0,y}(m)\,dy$$
$$+ \sum_{n,m\geq 2} e^{-\theta(n+m)} \int_{\mathbb{R}}\int_{\mathbb{R}} \widetilde{f}_{x,y}(n)\widetilde{f}_{x,y}(m)\,dx\,dy \ . \quad (1.32)$$

The first term on the r.h.s.

$$\widetilde{f}_{x,0}(1)^2 = e^{-\beta x^2}\left(\frac{\nu(0)}{\lambda\nu(-x)}\right)^2 = \left(\frac{\nu(0)}{\lambda}\right)^2 \exp\left(-\frac{x^2}{2}\left[2\beta + \alpha - \sqrt{\alpha}\sqrt{\alpha+4\beta}\right]\right)$$

is integrable for our conditions (**AP**) on α and β, because [...] > 0 (cf. Calculation A.9). Let us consider $n \geq 2$. From (1.25) and Remark 1.13 we know that $\widetilde{f}_{0,y}(n) = \varphi_n^{(0,0)}(y,0)\mathbb{1}_{\{y\neq 0\}} \leq c_1/\sqrt{n}$ for some constant c_1. In addition

$$\int_{\mathbb{R}} \widetilde{f}_{0,y}(m)\,dy = \int_{\mathbb{R}} \varphi_m^{(0,0)}(y,0)\mathbb{1}_{\{y\neq 0\}}\,dy < c_2, \text{ for all } m \geq 2,$$

because $\varphi_m^{(0,0)}(.,.)$ is a Gaussian density and $(\Sigma_m^{-1})_{1,1} \to c > 0$, as $m \to \infty$. Moreover for all $x,y \in \mathbb{R}$, $\varphi_m^{(0,0)}(x,y)$ is decreasing in m. So the second term in (1.32) is all right. For the last term observe that again from (1.25) and Remark 1.13 we know $\widetilde{f}_{x,y}(n) = \varphi_n^{(-x,0)}(y,0)\mathbb{1}_{\{y\neq 0\}} \leq c_1/\sqrt{n}$. Furthermore from proposition 1.12 and (1.21)

$$\widetilde{f}_{x,y}(m) = \varphi_m^{(-x,0)}(y,0)\mathbb{1}_{\{y\neq 0\}} = \varphi_m^{(0,0)}\left(y - \mu_{m-1}^{\alpha,\beta}(-x,0), -\mu_m^{\alpha,\beta}(-x,0)\right)\mathbb{1}_{\{y\neq 0\}}$$

and because $\varphi_m^{(0,0)}(.,.)$ is a (Gaussian) probability density, we have by (1.12) and (1.13)

$$\int_{\mathbb{R}}\int_{\mathbb{R}} \widetilde{f}_{x,y}(m)\,dx\,dy = \int_{\mathbb{R}}\int_{\mathbb{R}} \varphi_m^{(0,0)}\left(y - [x(\hat{c}_1+\hat{c}_2)+\hat{c}_3], -[x(\hat{c}_1-\hat{c}_2)+\hat{c}_3]\right)\mathbb{1}_{\{y\neq 0\}}\,dx\,dy$$
$$= \frac{1}{|\hat{c}_1-\hat{c}_2|}$$
$$= \left|\frac{2\sqrt{\alpha}}{\sqrt{\alpha}-\sqrt{\alpha+4\beta} - \left(\frac{\alpha}{\beta}\right)^m(\sqrt{\alpha}-\sqrt{\alpha+4\beta})}\right|$$
$$\leq \text{const.} \quad , \text{ for all } m \in \mathbb{N}$$

with appropriate values $\hat{c}_1, \hat{c}_2, \hat{c}_3$ from $\mu_{m-1}^{\alpha,\beta}$ and $\mu_m^{\alpha,\beta}$, so we are done. $\square$

1.6.2 Zerner's theorem and proof of Proposition 1.14

In particular, by the last Lemma we have shown the compactness of B^θ on $L^2(\mathbb{R}, d\mu)$. Thus we can apply an infinite dimensional Perron-Frobenius theorem of Zerner, cf. Appendix A.1. For this purpose let $\theta > 0$ and $\delta(\theta) \in (0,\infty)$ be the spectral radius of the operator B^θ. By Zerner's Theorem $\delta(\theta)$ is an isolated and simple eigenvalue of B^θ, therefore following analogous to [10], the $\delta(.)$ is strictly decreasing on $(0,\infty)$. Moreover, for $\theta > 0$ the state $\{0\}$ is a proper atom:

$$B^\theta_{0,\{0\}} = \sum_{n\in\mathbb{N}} e^{-\theta n} \widetilde{F}_{0,\{0\}}(n) = \sum_{n\in\mathbb{N}} e^{-\theta n} \int_{\{0\}} \widetilde{f}_{0,y}(n)\, \mu(dy) = \frac{1}{\lambda} e^{-\theta} > 0$$

and so by [25] chapter 4.2 also a small set of $B^\theta_{.,.}$. Hence, by [25] chapter 3.2, $\delta(.)$ can be represented in the variational formula

$$\delta(\theta) = \inf\left\{ \varrho > 0 \;\middle|\; \sum_{n=0}^\infty \varrho^{-n} \left(B^\theta\right)^{\circ n}_{0,\{0\}} < \infty \right\}. \qquad (1.33)$$

A very important thing is the behavior of the spectral radius $\delta(.)$ close to zero. Here we have to be careful, because $\delta(0)$ could possibly not exist and indeed

Proposition 1.17 *The "spectral radius" at the origin is $\delta(0) = \infty$.*

Proof We will prove that already the two-fold composition of $B^\theta_{x,dy}$ with itself diverges when $\theta \searrow 0$, i.e.

$$\left(B^\theta\right)^{\circ 2}_{0,\{0\}} \nearrow \infty \quad,\text{ when } \theta \searrow 0 \,.$$

Then from the variational formula it would follow that $\delta(\theta) \nearrow \infty$ for $\theta \searrow 0$.
First of all we have

$$B^\theta_{z,\{0\}} = \sum_{n\in\mathbb{N}} e^{-\theta n} \widetilde{F}_{z,\{0\}}(n) = \sum_{n\in\mathbb{N}} e^{-\theta n} \int_{\{0\}} \widetilde{f}_{z,y}(n)\, \mu(dy) = \frac{1}{\lambda} e^{-\theta} e^{-\beta z^2/2}$$

and

$$B^\theta_{0,dz} = \widetilde{f}_{0,z}(1)\, \delta_0(dz) + \sum_{n=2}^\infty e^{-\theta n} \widetilde{f}_{0,z}(n)\, dz \,.$$

Therefore

$$\left(B^\theta\right)^{\circ 2}_{0,\{0\}} = \int_{z\in\mathbb{R}} B^\theta_{0,dz} B^\theta_{z,\{0\}} = \frac{1}{\lambda} e^{-\theta} \left(\int_{z\in\mathbb{R}} e^{-\beta z^2/2} \widetilde{f}_{0,z}(1)\, \delta_0(dz) + \sum_{n=2}^\infty e^{-\theta n} \int_{z\in\mathbb{R}} e^{-\beta z^2/2} \widetilde{f}_{0,z}(n)\, dz \right)$$

$$= \frac{1}{\lambda^2} e^{-\theta} + \frac{1}{\lambda} e^{-\theta} \sum_{n=2}^\infty e^{-\theta n} \int_{z\in\mathbb{R}} e^{-\beta z^2/2} \widetilde{f}_{0,z}(n)\, dz \qquad (1.34)$$

Now we use (1.25) and 1.21 to write for $n \geq 2$

$$\widetilde{f}_{0,z}(n) = \frac{1}{2\pi\sqrt{\det \Sigma_n}} \exp\left\{ -\frac{z^2}{2} \left(\Sigma_n^{-1}\right)_{1,1} \right\}.$$

From 3.35 we know that there exists an $c_1 > 0$, such that $\left(\Sigma_n^{-1}\right)_{1,1} \leq c_1$ for all $n\in\mathbb{N}$ and by Proposition 1.12 we can bound from below

$$\int_{z\in\mathbb{R}} e^{-\beta z^2/2} \widetilde{f}_{0,z}(n)\, dz = \frac{1}{2\pi\sqrt{\det \Sigma_n}} \int_{z\in\mathbb{R}} \exp\left\{ -\frac{z^2}{2} \left(\beta + \left(\Sigma_n^{-1}\right)_{1,1} \right) \right\} dz$$

$$\geq \frac{1}{2\pi\sqrt{\det \Sigma_n}} \int_{z\in\mathbb{R}} \exp\left\{ -\frac{z^2}{2} (\beta + c_1) \right\} dz \geq \frac{c}{\sqrt{n}} \,.$$

1.6 Accurate determination of F_s

Now clearly by (1.34) we have

$$\lim_{\theta \searrow 0} \left(B^\theta\right)^{\circ 2}_{0,\{0\}} \geq \lim_{\theta \searrow 0} \left(\frac{1}{\lambda^2} e^{-\theta} + \frac{1}{\lambda} e^{-\theta} \sum_{n=2}^{\infty} e^{-\theta n} \frac{c}{\sqrt{n}} \right) = \infty \ .$$

□

Now we know indeed that $\delta(\theta) \nearrow \infty$ for $\theta \searrow 0$. With that we define the inverse $\delta^{-1}(.)$ on $(0,\infty)$ and set $\varepsilon_c := 0$ and

$$F_s(\varepsilon_c) := 0 \ , \qquad F_s(\varepsilon) := \delta^{-1}(1/\varepsilon) \quad , \text{ for } \varepsilon > 0 \ . \tag{1.35}$$

It is not a coincidence that the notation of F_s is close to that of the free energy, we will see later that both are even equal.

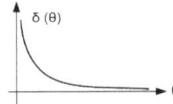

Figure 1.6: A sketch of the spectral radius $\delta(.)$. It is strcitly decreasing with $\lim_{\theta \searrow 0} \delta(\theta) = \infty$ and $\lim_{\theta \to \infty} \delta(\theta) = 0$.

For $\varepsilon > 0$ we consider the operator $B^{F_s(\varepsilon)}$ with the spectral radius $\delta(F_s(\varepsilon)) = 1/\varepsilon$. The kernel of $B^{F_s(\varepsilon)}$ is strictly positive, so Zerner's theorem A.1 ensures the existence of the right and left Perron-Frobenius eigenfunctions $\nu_\varepsilon(.), w_\varepsilon(.) \in L^2(\mathbb{R}, d\mu)$, such that $\nu_\varepsilon(x), w_\varepsilon(x) > 0$ for μ-a.e. $x \in \mathbb{R}$ and

$$\int_{y\in\mathbb{R}} B^{F_s(\varepsilon)}_{x,dy} \nu_\varepsilon(y) = \frac{1}{\varepsilon} \nu_\varepsilon(x) \quad , \quad \int_{x\in\mathbb{R}} w_\varepsilon(x) B^{F_s(\varepsilon)}_{x,dy} \mu(dx) = \frac{1}{\varepsilon} w_\varepsilon(y) \, \mu(dy) \ . \tag{1.36}$$

From this one even sees that $\nu_\varepsilon(x), w_\varepsilon(x) > 0$ for all $x \in \mathbb{R}$.

Proof of Proposition 1.14

We were not very precise in using the same notation for ε_c, F_s and ν_ε like the one in Proposition 1.14, since it is yet not clear if they satisfy what we would like to have. Nevertheless, the lines above tell us that the only remaining thing about those candidates is to prove (1.27), but using (1.36) this is indeed true

$$\int_{y\in\mathbb{R}} \sum_{n\in\mathbb{N}} K^\varepsilon_{x,dy}(n) = \frac{\varepsilon}{\nu_\varepsilon(x)} \int_{y\in\mathbb{R}} \left(\sum_{n\in\mathbb{N}} \widetilde{F}_{x,dy}(n) e^{-F_s(\varepsilon)n} \right) \nu_\varepsilon(y)$$

$$= \frac{\varepsilon}{\nu_\varepsilon(x)} \int_{y\in\mathbb{R}} B^{F_s(\varepsilon)}_{x,dy} \nu_\varepsilon(y) = 1 \ .$$

□

Remark 1.18 *According to (1.28) and Proposition 1.14, the process $(J_i)_{i\in\mathbb{Z}^+}$ is a Markov chain on $\mathbb{R}$. The chain starts in $J_0 = 0$ and has the transition kernel*

$$\mathcal{P}_\varepsilon(J_{i+1} \in dy \,|\, J_i = x) = \sum_{n\in\mathbb{N}} K^\varepsilon_{x,dy}(n) =: D^\varepsilon_{x,dy} \;.$$

The left and right eigenfunctions are defined up to multiplicative constant, so we can assume from now on that $\langle \nu_\varepsilon, w_\varepsilon\rangle_{L^2(\mathbb{R},d\mu)} = \int_\mathbb{R} \nu_\varepsilon w_\varepsilon \, d\mu = 1$. This means $\kappa_\varepsilon(dx) := \nu_\varepsilon(x)w_\varepsilon(x)\mu(dx)$ is a probability measure on $\mathcal{B}(\mathbb{R})$. Due to (1.36), if $\varepsilon > 0$ then κ_ε is invariant for $D^\varepsilon_{x,dy}$:

$$\int_{x\in\mathbb{R}} D^\varepsilon_{x,dy}\,\kappa_\varepsilon(dx) = \int_{x\in\mathbb{R}} \left(\sum_{n\in\mathbb{N}} \widetilde{F}_{x,dy}(n)e^{-F_s(\varepsilon)n}\right) \frac{\varepsilon}{\nu_\varepsilon(x)}\nu_\varepsilon(y)\nu_\varepsilon(x)w_\varepsilon(x)\mu(dx)$$

$$= \varepsilon\nu_\varepsilon(y)\int_{x\in\mathbb{R}} B^{F_s(\varepsilon)}_{x,dy}\,w_\varepsilon(x)\mu(dx) = \kappa_\varepsilon(dy)\;.$$

Therefore $(J_i)_{i\in\mathbb{Z}^+}$ is a positive recurrent Markov chain under $\mathcal{P}_\varepsilon$, if $\varepsilon > 0$, cf. [25].

1.7 Identification of the free energy and proof of Thm. 1.1

In this section we will prove the localization-delocalization result, which was stated in Theorem 1.1. In particular we will see the connection of previous results to the free energy defined in (1.2).

1.7.1 The double-contact process

We have already seen that $\{\tau_i\}_{i\in\mathbb{Z}^+}$ is a Markov renewal process. In what follows we need a "sub-process" of $\{\tau_i\}_{i\in\mathbb{Z}^+}$, which will be an ordinary (and non Markov) renewal process. Namely, we define the double-contact process $\{\eta_i\}_{i\in\mathbb{Z}^+}$ by

$$\eta_0 := 0 \quad, \quad \eta_{i+1} := \inf\{k > \eta_i \,|\, \varphi_{k-1} = \varphi_k = 0\} \tag{1.37}$$

and the index-process of returns to zero of $\{J_i\}_{i\in\mathbb{Z}^+}$:

$$\zeta_0 := 0 \quad, \quad \zeta_{i+1} := \inf\{k > \zeta_i \,|\, J_k = 0\}\;, \tag{1.38}$$

which we will also need later on in chapter 5.

1.7 Identification of the free energy and proof of Thm. 1.1

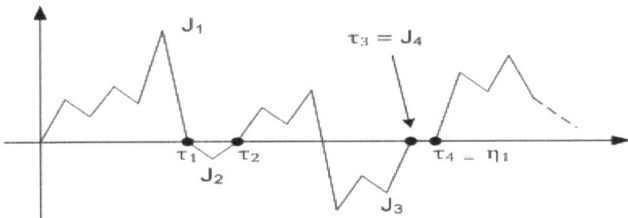

Figure 1.7: A sketch of the contact process $\{\varphi_i\}_{i\in\mathbb{Z}^+}$, the heights directly before contacts $\{J_i\}_{i\in\mathbb{Z}^+}$, the double contacts $\{\eta_i\}_{i\in\mathbb{Z}^+}$ and the index-process $\{\zeta_i\}_{i\in\mathbb{Z}^+}$ of returns to zero of $\{J_i\}_{i\in\mathbb{Z}^+}$ under $\mathbb{P}_{\varepsilon,N}$. For instance we have the following relations: $\zeta_1 = 4, 0 = J_{\zeta_1} = \varphi_{\tau_3} = \varphi_{\tau_{\zeta_1}-1}$ and $\tau_{\zeta_1} = \eta_1$.

Because of the special structure of the transition kernel (1.28) and the remark 1.18 the following proposition of [10] applies.

Proposition 1.19 *For each $\varepsilon > 0$ under $\mathcal{P}_\varepsilon$ the double-contact process $(\eta_i)_{i\in\mathbb{Z}^+}$ is a non terminating renewal process.*

1.7.2 Proof of Theorem 1.1

We will first show that the pinning model displays a trivial phase transition, meaning that indeed $\varepsilon_c = 0$. To obtain this, it remains to show that the expression F_s, defined in (1.35), for all $\varepsilon \geq 0$ indeed coincides with the free energy from Definition (1.2), i.e.

$$F(\varepsilon) = \lim_{N\to\infty} \frac{1}{N} \log\left(\frac{\mathcal{Z}_{\varepsilon,N}}{\mathcal{Z}_{0,N}}\right).$$

Now, with the help of (1.31) we can write for $\varepsilon > 0$

$$\frac{\mathcal{Z}_{\varepsilon,N}}{\mathcal{Z}_{0,N}} = \frac{e^{F_s(\varepsilon)(N+1)}}{\varepsilon^2 \mathcal{Z}_{0,N}} \lambda^{N+1} \mathcal{P}_\varepsilon(\mathcal{A}_N)$$

and so

$$\frac{1}{N}\log\left(\frac{\mathcal{Z}_{\varepsilon,N}}{\mathcal{Z}_{0,N}}\right) = \frac{N+1}{N}F_s(\varepsilon) + \frac{N+1}{N}\log\lambda + \frac{1}{N}\log\mathcal{P}_\varepsilon(\mathcal{A}_N) - \frac{2}{N}\log\varepsilon - \frac{1}{N}\log\mathcal{Z}_{0,N}.$$
(1.39)

Due to Proposition 1.11, in the limit $N \to \infty$ we can neglect the second and last term on the right hand side. In view of the Definition (1.35) of F_s there are two cases to distinguish between, namely $\varepsilon = 0$ and $\varepsilon > 0$.

For $\varepsilon = 0$ it is trivial, because $F(0) = 0$, as is obvious from the definition and $F_s(0)$ was defined to be 0 in (1.35). We turn to the second case $\varepsilon > 0$, in which, by definition (1.35), $F_s(\varepsilon) > 0$. Considering (1.39), to complete the identification of the free energy it remains to check that one has

$$\lim_{N \to \infty} \frac{1}{N} \log \mathcal{P}_\varepsilon(\mathcal{A}_N) = 0 . \tag{1.40}$$

The set $\mathcal{A}_N$, defined in proposition 1.15, can be written as $\mathcal{A}_N = \{\exists j \geq 0 \,|\, \eta_j = N+1\}$. It is known that (1.40) is true for any non-terminating aperiodic renewal process, cf. [18] Theorem A.3. However $(\eta_i)_{i \in \mathbb{Z}^+}$ is aperiodic, because

$$\mathcal{P}_\varepsilon(\eta_1 = 1) = \mathcal{P}_\varepsilon((\tau_1, J_1) \in (\{1\}, \{0\}) \,|\, (\tau_0, J_0) = (0,0)) = K^\varepsilon_{0,\{0\}}(1) = \frac{\varepsilon}{\lambda} e^{-F_s(\varepsilon)} > 0$$

and due to proposition 1.19 it is a non-terminating renewal process under $\mathcal{P}_\varepsilon$ for $\varepsilon > 0$. Altogether we have shown

$$F_s(\varepsilon) = F(\varepsilon) = \lim_{N \to \infty} \frac{1}{N} \log \left(\frac{\mathcal{Z}^w_{\varepsilon, N}}{\mathcal{Z}^w_{0, N}} \right) \quad , \varepsilon \geq 0 .$$

We have already studied the property of analyticity in the localized regime $\mathcal{L}$ before we defined F_s. Finally what is left is the asymptotic behavior of the free energy $F(\varepsilon)$ as $\varepsilon \to \infty$. The idea is to use a sandwich argument and therefore first of all by the Definition in Lemma 1.16 we have

$$e^{-\theta} \widetilde{F}_{x,dy}(1) \leq B^\theta_{x,dy} \leq e^{-\theta} e^{\theta_0} B^{\theta_0}_{x,dy} \quad , \text{ for all } \theta \geq \theta_0 , \tag{1.41}$$

for an arbitrarily chosen $\theta_0 > 0$. The last inequality is true because

$$B^\theta_{x,dy} = e^{-\theta} \sum_{n \in \mathbb{N}} e^{-\theta(n-1)} \widetilde{F}_{x,dy}(n) \leq e^{-\theta} \sum_{n \in \mathbb{N}} e^{-\theta_0(n-1)} \widetilde{F}_{x,dy}(n) = e^{-\theta} e^{\theta_0} B^{\theta_0}_{x,dy} .$$

Further on one can consider their corresponding integral operators on $L^2(\mathbb{R}, d\mu)$, e.g. $(\widetilde{\mathcal{F}} h)(x) := \int \widetilde{F}_{x,dy}(1) h(y)$. In particular it is by (1.24) for $h(x) = e^{-V_2(x)}/\nu(-x)$:

$$(\widetilde{\mathcal{F}} h)(x) = \int \widetilde{F}_{x,y}(1) h(y) \, \delta_0(dy) = \frac{1}{\lambda} h(x) .$$

Now the same inequalities as in (1.41) have to be valid for the spectral radius of B^θ and B^{θ_0}, this means

$$\frac{1}{\lambda} e^{-\theta} \leq \delta(\theta) \leq e^{-\theta} e^{\theta_0} \delta(\theta_0) \quad , \text{ for all } \theta \geq \theta_0 .$$

Now we set $\theta := (\delta)^{-1}(1/\varepsilon) \geq \theta_0$, i.e. $F(\varepsilon) \geq \theta_0$. This means, for all $\varepsilon \geq \varepsilon_0 := 1/\delta(\theta_0)$ we obtain

$$\log(\varepsilon/\lambda) \leq F(\varepsilon) \leq \log\left(\varepsilon e^{\theta_0} \delta(\theta_0)\right) ,$$

which implies the asymptotic behavior in Theorem 1.1.

1.8 Alternative proof of trivial phase transition

In the last section of this chapter we want to show an alternative way how to prove a trivial phase transition ($\varepsilon_c = 0$) for the pinning model (1.1). For this purpose let us write the product measure in a specific way.

Lemma 1.20 *The product measure in* (1.1) *has the following expansion*

$$\prod_{i=1}^{N-1}(\varepsilon\delta_0(d\varphi_i) + d\varphi_i) = \prod_{i=1}^{N-1} d\varphi_i + \varepsilon^2 \sum_{i=1}^{N-2} \left(\prod_{j=1}^{i-1} d\varphi_j\right) \delta_0(d\varphi_i)\delta_0(d\varphi_{i+1}) \prod_{k=i+2}^{N-1}(\varepsilon\delta_0(d\varphi_k) + d\varphi_k)$$

$$+ \varepsilon \sum_{i=1}^{N-2} \left(\prod_{j=1}^{i-1} d\varphi_j\right) \delta_0(d\varphi_i)d\varphi_{i+1} \prod_{k=i+2}^{N-1}(\varepsilon\delta_0(d\varphi_k) + d\varphi_k)$$

$$+ \varepsilon \left(\prod_{j=1}^{N-2} d\varphi_j\right) \delta_0(d\varphi_{N-1}) \ .$$

Proof Let $M := \{1, ..., N-1\}$, then the standard expansion can be written like in (1.22)

$$\prod_{i=1}^{N-1}(\varepsilon\delta_0(d\varphi_i) + d\varphi_i) = \sum_{A \subseteq M} \varepsilon^{|A|} \left(\prod_{i \in A} \delta_0(d\varphi_i)\right) \left(\prod_{j \in M \setminus A} d\varphi_j\right) \ . \tag{1.42}$$

Denote by $\mathcal{P}(M)$ the power set of M and

$$M_1 = \bigcup_{i \in \{1,...,N-2\}}^{\cdot} M_1^i \ , \quad M_1^i = \{A \subseteq M \mid \min\{k | k \in A\} = i \text{ and } i+1 \in A\} \ ,$$

$$M_2 = \bigcup_{i \in \{1,...,N-2\}}^{\cdot} M_2^i \ , \quad M_2^i = \{A \subseteq M \mid \min\{k | k \in A\} = i \text{ and } i+1 \notin A\} \ ,$$

$$M_3 = \{A \subseteq M \mid \min\{k | k \in A\} = N-1\} = \{N-1\} \quad , \quad M_4 = \emptyset \ .$$

Therefore one easily sees that the M_i's are disjoint. Furthermore for $i \in \{1, ..., N-2\}$ the cardinality of the power set $\mathcal{P}\{i+2, ..., N-1\}$ is 2^{N-2-i} and so M_1^i and M_2^i each consist of 2^{N-2-i} elements. Altogether the cardinality

$$|M_1 \dot\cup M_2 \dot\cup M_3 \dot\cup M_4| = 2 \sum_{i=1}^{N-2} 2^{N-2-i} + 2 = 2 \cdot 2^{N-2} \left(\frac{1 - 2^{-(N-1)}}{1 - 2^{-1}} - 1\right) + 2 = 2^{N-1}$$

is equal to the cardinality of $\mathcal{P}(M)$ and so we have $\mathcal{P}(M) = M_1 \dot\cup M_2 \dot\cup M_3 \dot\cup M_4$. Due to the disjointness of the M_i's it is obvious from (1.42) that the r.h.s. in the lemma can be obtained by summing over the sets M_4, M_1, M_2 and M_3. □

Now, let us recall the partition function for the model

$$\mathcal{Z}_{\varepsilon,N} = \int_{\mathbb{R}^{N-1}} e^{-\mathcal{H}_{[-1,N]}(\varphi)} \prod_{i=1}^{N-1}(\varepsilon\delta_0(d\varphi_i) + d\varphi_i)$$

where we have $\varphi_{-1} = \varphi_0 = \varphi_N = \varphi_{N+1} = 0$ and

$$\mathcal{H}_{[-1,N+1]}(\varphi) = \alpha \sum_{i=1}^{N+1} \frac{1}{2}(\nabla \varphi_i)^2 + \beta \sum_{i=0}^{N} \frac{1}{2}(\Delta \varphi_i)^2 \ .$$

Therefore by the last Lemma we can "decouple" the partition function and obtain something like a "renewal inequality" for $0 \leq \varepsilon \leq 1$

$$\mathcal{Z}_{\varepsilon,N} \geq \mathcal{Z}_{0,N} + \varepsilon^2 \sum_{i=1}^{N-1} \mathcal{Z}_{0,i} \mathcal{Z}_{\varepsilon, N-i-1} \ , \tag{1.43}$$

where $\mathcal{Z}_{\varepsilon,0} := 1$. Furthermore by Proposition 1.11 and its proof

$$\mathcal{Z}_{0,N} = \frac{\lambda^{N-1}}{\sqrt{N-1}} \kappa(N), \quad \kappa(N) = \left(\frac{(N-1)\sigma_+^{N-1}}{\det(\alpha A_{N-1} + \beta B_{N-1})} \right)^{1/2} , \tag{1.44}$$

where by (1.11) $\kappa(N) \sim (c_2 + o(1))^{-1/2}$ as $N \to \infty$ and λ is the spectral radius from Proposition 1.5. We can state the following

Proposition 1.21 *For every $\varepsilon > 0$ we have $F(\varepsilon) > 0$. More precisely the following lower bound holds*

$$F(\varepsilon) \geq -\log(x(\varepsilon)) \tag{1.45}$$

where $x = x(\varepsilon) \in (0,1)$ is the unique solution of

$$g(x) := x^2 + \sum_{n=2}^{\infty} \frac{x^{n+1}}{\sqrt{n-1}} \kappa(n) = \frac{\lambda^2}{\varepsilon^2} \ . \tag{1.46}$$

Proof First observe that $g(0) = 0$ and $g(x) \nearrow \infty$ if $x \nearrow 1$, also is g strictly increasing in x. Fix any $\varepsilon > 0 =: \varepsilon_c$ and take $x := x(\varepsilon) \in (0,1)$ as the corresponding unique solution of $g(x) = \lambda^2/\varepsilon^2$. We extend here the idea of the proof in [7]. For $n \geq 1$ we define

$$u_n := \frac{x^n \mathcal{Z}_{\varepsilon,n}}{\lambda^{n-1}} \quad , \quad a_n := \frac{\varepsilon^2 x^{n+1} \mathcal{Z}_{0,n}}{\lambda^{n+1}} \quad , \quad b_n := \frac{a_n \lambda^2}{\varepsilon^2 x} \ .$$

We set $a_0 := b_0 := 0$ and $u_0 := \lambda$. In particular $a_1 = \varepsilon^2 x^2 \lambda^{-2}$, $b_1 = x$ and due to (1.44) for $n \geq 2$

$$a_n = \frac{\varepsilon^2}{\lambda^2} \frac{x^{n+1}}{\sqrt{n-1}} \kappa(n) \ .$$

Because of (1.43) we get

$$u_n \geq b_n + \sum_{i=0}^{n-1} u_{n-i-1} = b_n + (a_{-1} u_n + a_0 u_{n-1} + \ldots + a_{n-1} u_0) \quad , \quad a_{-1} := 0$$

$$= b_n + (\tilde{a}_0 u_n + \ldots + \tilde{a}_n u_0) \tag{1.47}$$

where we set $\tilde{a}_i := a_{i-1}$ for $i = 0, 1, 2, \ldots$. Now let $\tilde{u}_n$ be defined by

$$\tilde{u}_n = b_n + \sum_{i=0}^{n} \tilde{a}_i \tilde{u}_{n-i} \quad , \quad \tilde{u}_0 = \lambda \ . \tag{1.48}$$

1.8 Alternative proof of trivial phase transition

By definition of a_n and the choice of $x = x(\varepsilon)$ we know that

$$\sum_{n=0}^{\infty} \tilde{a}_n = \sum_{n=0}^{\infty} a_{n-1} = \frac{\varepsilon^2}{\lambda^2} x^2 + \frac{\varepsilon^2}{\lambda^2} \sum_{n=2}^{\infty} \frac{x^{n+1}}{\sqrt{n-1}} \kappa(n) = \frac{\varepsilon^2}{\lambda^2} g(x(\varepsilon)) = 1 \;.$$

But this is enough to apply a theorem from the renewal theory (cf. [15] chap.XIII) on the sequence $\tilde{u}_n$ defined by (1.48) and we get

$$\lim_{n \to \infty} \tilde{u}_n = \frac{B}{A}$$

where

$$B := \sum_{n=0}^{\infty} b_n = x + \sum_{n=2}^{\infty} \frac{x^n}{\sqrt{n-1}} \kappa(n) = x g(x) = x \frac{\lambda^2}{\varepsilon^2}$$

and

$$A := \sum_{n=0}^{\infty} n \tilde{a}_n = \frac{\varepsilon^2}{\lambda^2} \left[2x^2 + \sum_{n=2}^{\infty} \frac{(n+1)x^{n+1}}{\sqrt{n-1}} \kappa(n) \right] = \frac{\varepsilon^2}{\lambda^2} x g'(x) \;.$$

Therefore the limit is

$$\lim_{n \to \infty} \tilde{u}_n = \frac{\lambda^4}{\varepsilon^4 g'(x)} \tag{1.49}$$

The next what we show is that for every sequence u_n, which fulfills (1.47) and $u_0 = \lambda$, it holds

$$\lim_{n \to \infty} u_n \geq \lim_{n \to \infty} \tilde{u}_n \;. \tag{1.50}$$

Set $\hat{u}_n := u_n - \tilde{u}_n$ for $n \geq 0$, then by (1.47) and (1.48) we have

$$\hat{u}_n \geq \sum_{i=0}^{n} \tilde{a}_i \hat{u}_{n-i} \quad , \quad \hat{u}_0 = 0 \;. \tag{1.51}$$

We claim that $\hat{u}_n \geq 0$ for every $n \in \mathbb{N}_0$. For $n = 0$ it is $\hat{u}_0 = \lambda - \lambda = 0$. Let $\hat{u}_0, \hat{u}_1, ..., \hat{u}_n \geq 0$ then by (1.51) and induction it follows

$$\hat{u}_{n+1} \geq \sum_{i=0}^{n+1} \tilde{a}_i \hat{u}_{n+1-i} = \tilde{a}_1 \hat{u}_n + ... + \tilde{a}_{n+1} \hat{u}_0 \geq 0$$

due to the fact, that $\tilde{a}_0 = \hat{u}_0 = 0$ and $\tilde{a}_1, ..., \tilde{a}_n \geq 0$. So we get $\lim_{n \to \infty} \hat{u}_n \geq 0$ and as we know from (1.49) that $\lim_{n \to \infty} \tilde{u}_n$ exists, we get finally (1.50) and

$$\lim_{n \to \infty} u_n \geq \frac{\lambda^4}{\varepsilon^4 g'(r)} \;.$$

By the definition of u_n

$$\mathcal{Z}_{\varepsilon,N} \succcurlyeq \frac{\lambda^{N+3}}{x^N \varepsilon^4 g'(x)} \;.$$

Dividing by $\mathcal{Z}_{0,N}$ and using (1.44) we have

$$\frac{\mathcal{Z}_{\varepsilon,N}}{\mathcal{Z}_{0,N}} \succcurlyeq \frac{\lambda^4 (c_2 + o(1))^{1/2}}{\varepsilon^4 g'(x)} \frac{\sqrt{N-1}}{x^N}$$

and therefore

$$F(\varepsilon) = \lim_{N \to \infty} \frac{1}{N} \log\left(\frac{\mathcal{Z}_{\varepsilon,N}}{\mathcal{Z}_{0,N}}\right) \geq -\log(x) \quad , \quad x = x(\varepsilon) \;.$$

$\square$

1.9 Modification of the pinning model

In this chapter we have seen that the pinning model displays a localization behavior as soon as an arbitrary positive pinning-strength is present. This behavior prevails even if we choose very strong potentials for the Laplacian interaction, cf. chapter 2. In this section we will artificially force the pinning model to display a proper phase transition. Since pinning by an constant $\varepsilon > 0$ does not reflect this behavior, we could for example pin by weaker strength. In the following we take a pinning-strength according to the absence-time from 0. Consider the same model for $\varphi = \{\varphi_1, ..., \varphi_{N-1}\}$ as in 1.1, but with modified pinning

$$\mathbb{P}_{\varepsilon,N}(d\varphi) := \frac{\exp(-\mathcal{H}_{[-1,N+1]}(\varphi))}{\mathcal{Z}_{\varepsilon,N}} \prod_{i=1}^{N-1} (a_i(\varphi)\varepsilon\delta_0(d\varphi_i) + d\varphi_i) \; ,$$

where

$$a_i(\varphi) := \begin{cases} a(i)a(N - \tau_{\ell_N - 1}) & , i = \tau_1 \\ a(\tau_j - \tau_{j-1}) & , i = \tau_j, j = 2, ..., \ell_N - 1 \\ 1 & , \text{otherwise} \; . \end{cases}$$

Here $\{a(n)\}_{n\in\mathbb{N}}$ is a positive sequence with $a(n) \to 0$, $n \to \infty$. Although artificial, one can ask what kind of situation does this model describe. Well, apparently the pinning-strength is smaller when the last contact is far way. This behavior could be interpreted in our case in the way that "longer pieces" of this polymer are harder to localize. In any way, we will see that the behavior of the contact process is immediately influenced. How can we see that? Similarly to the pinning model (1.1) we can consider the distribution of the contacts in the following way. Recall the number of contacts to the defect line $\ell_N = \#\{i \in \{1, ..., N\} \mid \varphi_i = 0\}$. Now take for fixed $k \geq 2$ a time-partition $(t_i)_{i=1,...,k} \in \mathbb{N}$ with $0 < t_1 < \cdots < t_{k-1} < t_k := N$. Moreover suppose the contacts are only at $\tau_i = t_i$, $i = 1, ..., k - 1$ then

$$\mathbb{P}_{\varepsilon,N}(\ell_N = k, \tau_i = t_i, J_i \in dy_i, \; i = 1, ..., k)$$
$$= \frac{\varepsilon^{k-1}}{\mathcal{Z}_{\varepsilon,N}} a(t_1)a(t_2 - t_1) \cdots a(N - t_{k-1}) F_{0,dy_1}(t_1) F_{y_1,dy_2}(t_2 - t_1) \cdots F_{y_{k-1},dy_k}(N - t_{k-1}) F_{y_k,\{0\}}(1) \; .$$
(1.52)

This can be again modified by the following semi Markov (sub)-kernel

$$K^{\varepsilon}_{x,dy}(n) := \varepsilon \widetilde{F}_{x,dy}(n) a(n) e^{-F(\varepsilon)n} \frac{\nu_\varepsilon(y)}{\nu_\varepsilon(x)}$$

to obtain

$$\mathcal{Z}_{\varepsilon,N} = \frac{e^{F(\varepsilon)(N+1)}}{\varepsilon^2 a(1)} \lambda^{N+1} \mathcal{P}_\varepsilon(\mathcal{A}_N) \; .$$

We don't want to go into detail, but confering chapter 1 and 3 one can show that this model exhibits a (proper) phase transition if and only if

$$a(n) = o\left(n^{-1/2}\right) \; .$$

Therfore we can say that $1/\sqrt{n}$ is the critical decay for the substantial difference between trivial and non-trivial phase transition.

2 General pinning model

2.1 Introduction, model, results

2.1.1 Introduction

In this chapter we are going to generalize the Gaussian pinning model that was defined in (1.1). In the last chapter we have always observed a trivial phase transition, no matter how small the parameter in front of the gradient part was. Nevertheless one should remark that the potential was of quadratic form and maybe still too strong. Here the results of a Gaussian interaction-potential will be extended to more general interaction potentials. In this setting one could be inclined to expect a non-trivial phase transition $\varepsilon_c > 0$, taking a very strong Laplacian potential and at the same time a weak gradient potential. However it turns out that even very strong Laplacian potentials are not able to change the localization behavior. In other words, the number of contacts to the defect line growth linearly for arbitrarily small pinning parameters $\varepsilon > 0$.

2.1.2 The model

We consider the model determined by the following distribution on $\mathbb{R}^{N-1}$:

$$\mathbb{P}_{\varepsilon,N}(d\varphi) := \frac{\exp(-\mathcal{H}_{[-1,N+1]}(\varphi))}{\mathcal{Z}_{\varepsilon,N}} \prod_{i=1}^{N-1} (\varepsilon \delta_0(d\varphi_i) + d\varphi_i) , \qquad (2.1)$$

where $N \in \mathbb{N}$, $\varepsilon \geq 0$ denotes the pinning parameter and $\delta_0(.)$ the Dirac mass at zero. Furthermore $d\varphi_i$ is the Lebesgue measure on $\mathbb{R}$ and $\mathcal{Z}_{\varepsilon,N}$ is a normalization constant, called partition function, to turn $\mathbb{P}_{\varepsilon,N}$ into a probability measure. Since we do not expect a conflict in the notation, we will use the same symbols like in the previous chapter. We define the Hamiltonian

$$\mathcal{H}_{[-1,N+1]}(\varphi_{-1}, ..., \varphi_{N+1}) := \sum_{i=1}^{N+1} V_1(\nabla\varphi_i) + \sum_{i=0}^{N} V_2(\Delta\varphi_i) , \qquad (2.2)$$

and choose zero boundary conditions $\varphi_{-1} = \varphi_0 = \varphi_N = \varphi_{N+1} = 0$. Next we consider continuous interaction potentials $V_1, V_2 : \mathbb{R} \to \mathbb{R}$ with the following properties:

C1 symmetric potential V_1, i.e. $V_1(x) = V_1(-x)$ for all $x \in \mathbb{R}$

C2 $\exists M > 0 : V_2$ is decreasing on $(-\infty, M]$ and increasing on $[M, \infty)$

C3 $\int e^{-V_2(x)}\,dx < \infty$ and $e^{-V_1(x)}$ bounded

C4 $\int |x| e^{-V_1(x)}\,dx < \infty$ and e^{-V_2} is bounded .

Remark 2.1
A sufficient condition for **C4** would be a growth rate of at least $V_1(z) > (2+\delta)\ln|z|$ for $|z| \to \infty$ and $\delta > 0$. We don't know what happens if we abandon condition **C4**, since for instance, it is needed in Proposition 2.8.

2.1.3 The result

So far a trivial phase transition, meaning $\varepsilon_c = 0$, has been established in Gaussian case, i.e. for interaction potentials of the type

$$V_1(\eta) = \frac{\alpha}{2}\eta^2 \quad \text{and} \quad V_2(\eta) = \frac{\beta}{2}\eta^2 ,$$

where $\alpha, \beta > 0$, cf. Theorem 1.1. The extremal cases of pure gradient or Laplacian interaction have been already investigated in a non-Gaussian setting. We refer to [12] and [20] in the gradient case and to [10] for the Laplacian model. These models display substantially different localization behavior. While the gradient interaction shows a trivial phase transition, the Laplacian interaction leads to a "real" phase transition, i.e. $\varepsilon_c > 0$. Regarding those facts, a natural question arises: what happens, if we choose a weak interaction potential V_1 and a very strong potential V_2 ? We were able to prove the following

Theorem 2.2 (Localization for general potentials) *For general continuous interaction potentials with conditions **C1**-**C4** the pinning model displays a trivial phase-transition.*

Somewhat surprisingly we see that even arbitrarily strong Laplacian potentials V_2 have no impact on the localization, the model behaves just like the gradient model. On the other hand it underlines again the semi-flexible character of polymers with length beyond the persistence length, cf. section 0.1.

The rest of this chapter deals with the proof of the Theorem 2.2. We have divided it into four parts, where in the first part we deal with the free model ($\varepsilon = 0$) and show that it can be seen as a bridge of an integrated Markov chain. In the second part we obtain certain bounds for the density of the Markov chain and its sum. The third part will provide the construction of a "double-contact" process from the partition function $\mathcal{Z}_{\varepsilon,N}$. Finally in the last part we obtain a sufficient lower bound for the free energy. We remark that an explicit representation of the free energy similar to chapter 1 won't be given, since we don't have a local limit theorem for the density $\varphi_n^{(a,b)}(x,y)$.

2.2 Markovian description of the free model

With our conditions on the potential we are now able to construct a Markov chain, which will be the basis for further investigations. Consider the linear integral operator

$$(\mathcal{K}f)(x) := \int k(x,y) f(y)\,dy \quad , \text{ where } \quad k(x,y) = e^{-V_1(y) - V_2(y-x)} \ .$$

Of course, immediately one might and should ask on which space this operator is defined. The next Proposition will provide an answer.

Proposition 2.3 $\mathcal{K}$ *is compact on* $L^\infty(\mathbb{R}, ||.||_\infty)$.

Proof For the proof of compactness we will apply an useful result of [14], namely Theorem 5.1 and its Corollary (cf. also Theorem 4.7 in this thesis). With that and $B(0,R) = \{x \in \mathbb{R} : |x| < R\}$ it is sufficient to show that the following three conditions are fulfilled

(i) $\exists C > 0$ such that for almost all $x \in \mathbb{R}$ $\int_\mathbb{R} |k(x,y)|\,dy < C$

(ii) $\forall \varepsilon > 0$ $\exists R > 0$ such that for almost all $x \in \mathbb{R}$ $\int_{\mathbb{R} \setminus B(0,R)} |k(x,y)|\,dy < \varepsilon$

(iii) $\forall \varepsilon > 0$ $\exists \delta > 0$ such that for almost all $x \in \mathbb{R}$ and $|h| < \delta$

$$\int_\mathbb{R} |k(x, y+h) - k(x,y)|\,dy < \varepsilon \ .$$

Now condition (i) is easily seen by Hölder and the property **C3**, for all $x \in \mathbb{R}$:

$$\int_\mathbb{R} |k(x,y)|\,dy \leq ||e^{-V_1(.)}||_\infty \int_\mathbb{R} e^{-V_2(y)}\,dy =: C < \infty \ .$$

Next for (ii), let $\varepsilon > 0$ and choose $R \in \mathbb{R}$ such that $\int_{\mathbb{R} \setminus B(0,R)} e^{-2V_1(y)}\,dy < \varepsilon^2 / \int_\mathbb{R} e^{-2V_2(y)}\,dy$. Indeed, conditions **C3** and **C4** imply $\int_\mathbb{R} e^{-V_i(y)}\,dy < \infty$ for $i = 1, 2$ and

$$\int_\mathbb{R} e^{-2V_i(y)}\,dy \leq \int_\mathbb{R} e^{-V_i(y)} \mathbf{1}_{\{V_i \geq 0\}}\,dy + (\text{const.}) \cdot \text{Leb}(\{V_i < 0\}) \ ,$$

where the last term has to be finite since

$$\infty > \int_\mathbb{R} e^{-V_i(y)}\,dy \geq \int_\mathbb{R} e^{-V_i(y)} \mathbf{1}_{\{V_i < 0\}}\,dy \geq \text{Leb}(\{V_i < 0\}) \ .$$

Therefore by the Cauchy-Schwarz inequality we obtain

$$\int_{\mathbb{R} \setminus B(0,R)} |k(x,y)|\,dy \leq \left(\int_{\mathbb{R} \setminus B(0,R)} e^{-2V_1(y)}\,dy \right)^{1/2} \left(\int_\mathbb{R} e^{-2V_2(y-x)}\,dy \right)^{1/2} < \varepsilon \ . \quad (2.3)$$

Finally we prove (iii). Let again $\varepsilon > 0$ be fixed and choose $R \in \mathbb{R}$ like in proof of (ii). For every "small" $\delta > 0$, which will be specified later, and all $|h| < \delta$ by triangle-inequality

$$\int_{\mathbb{R}} |k(x, y+h) - k(x, y)| \, dy = \int_{\mathbb{R}\setminus B(0, R+\delta)} |k(x, y+h) - k(x, y)| \, dy$$
$$+ \int_{B(0, R+\delta)} |k(x, y+h) - k(x, y)| \, dy$$
$$\leq \int_{\mathbb{R}\setminus B(0, R+\delta)} k(x, y+h) \, dy + \int_{\mathbb{R}\setminus B(0, R+\delta)} k(x, y) \, dy + \int_{B(0, R+\delta)} |k(x, y+h) - k(x, y)| \, dy$$
$$< \frac{2}{3}\varepsilon + \int_{B(0, R+\delta)} |k(x, y+h) - k(x, y)| \, dy \, .$$

The last bound is due to the fact that for $|h| < \delta$ it is $\{y + h \mid y \in \mathbb{R}\setminus B(0, R+\delta)\} \subseteq \mathbb{R}\setminus B(0, R)$ and then we proceed like in (2.3). Next we set $g(x, y, h) := |k(x, y+h) - k(x, y)|$. To finish the proof we have to find a $\delta > 0$ such that for almost all $x \in \mathbb{R}$ and $|h| < \delta$

$$\int_{B(0, R+\delta)} g(x, y, h) \, dy < \frac{\varepsilon}{3} \, .$$

Now we are going to investigate the supremum over x. For all $\widetilde{R} > 0$

$$\sup_{x \in \mathbb{R}} \int_{B(0, R+\delta)} g(x, y, h) \, dy \leq \int_{-R-\delta}^{R+\delta} \left[\sup_{x: |y-x| \in B(0, \widetilde{R})} g(x, y, h) + \sup_{x: |y-x| \in \mathbb{R}\setminus B(0, \widetilde{R})} g(x, y, h) \right] dy$$
$$= \int_{-R-\delta}^{R+\delta} \sup_{z \in B(0, \widetilde{R})} \widetilde{g}(z, y, h) \, dy + \int_{-R-\delta}^{R+\delta} \sup_{z \in \mathbb{R}\setminus B(0, \widetilde{R})} \widetilde{g}(z, y, h) \, dy$$

(2.4)

where
$$\widetilde{g}(z, y, h) := \left| e^{-V_1(y+h) - V_2(z+h)} - e^{-V_1(y) - V_2(z)} \right| \, .$$

Since V_1, V_2 are continuous, for every $|h| < \delta$ the function $\widetilde{g}$ attains its maximum on $D := \overline{B(0, R+\delta)} \times \overline{B(0, \widetilde{R})}$, say at (z_h, y_h). On D is $\widetilde{g}$ uniformly continuous, therefore $\widetilde{g}(z_h, y_h, h) \to 0$ for $|h| \to 0$. This means we can choose a $\delta > 0$ such that for all $|h| < \delta$

$$\widetilde{g}(z_h, y_h, h) < \frac{\varepsilon}{6 \operatorname{Vol}(\overline{B(0, R+\delta)})} \, . \tag{2.5}$$

Next by triangle-inequality and **C4** for the second integral in (2.4) we have

$$\int_{-R-\delta}^{R+\delta} \sup_{z \in \mathbb{R}\setminus \overline{B(0, \widetilde{R})}} \left| e^{-V_1(y+h) - V_2(z+h)} - e^{-V_1(y) - V_2(z)} \right| dy$$
$$\leq \sup_{z \in \mathbb{R}\setminus \overline{B(0, \widetilde{R})}} e^{-V_2(z+h)} \int_{-R-\delta}^{R+\delta} e^{-V_1(y+h)} dy + \sup_{z \in \mathbb{R}\setminus \overline{B(0, \widetilde{R})}} e^{-V_2(z)} \int_{-R-\delta}^{R+\delta} e^{-V_1(y)} dy$$
$$\leq \text{const.} \sup_{z \in \mathbb{R}\setminus \overline{B(0, \widetilde{R}-\delta)}} e^{-V_2(z)} \, . \tag{2.6}$$

2.2 Markovian description of the free model

Because of **C2** and **C3** the last expression converges to 0, when $\tilde{R} \to \infty$. Therefore we can choose $\tilde{R} > 0$ such that (2.6) is smaller than $\varepsilon/6$. Collecting all together by (2.4) and (2.5) we have proven

$$\sup_{x \in \mathbb{R}} \int_{B(0,R+\delta)} g(x,y,h)\, dy < \frac{\varepsilon}{3} \ .$$

$\square$

The infinite dimensional Perron-Frobenius Theorem A.1 ensures now the existence of an isolated spectral radius $\lambda > 0$ of $\mathcal{K}$ and the corresponding left and right eigenfunctions $w \in L^1(\mathbb{R}, \|.\|_1)$, $\nu \in L^\infty(\mathbb{R}, \|.\|_\infty)$ with $0 < w(x)$ and $\nu(x) > 0$ a.s. . This can be extended to $w(x) > 0$ and $\nu(x) > 0$ for all $x \in \mathbb{R}$, due to the eigenfunction representations

$$\nu(x) = \frac{1}{\lambda} \int_{\mathbb{R}} k(x,y)\, \nu(y)\, dy \quad \text{and} \quad w(y) = \frac{1}{\lambda} \int_{\mathbb{R}} k(x,y)\, w(x)\, dx \ .$$

With that we are able to construct a Markov chain. For $a, b \in \mathbb{R}$ we consider a probability space $(\Omega, \mathcal{A}, \mathrm{P}^{(a,b)})$ and two processes $\{Y_i\}_{i \in \mathbb{Z}^+}$, $\{W_i\}_{i \in \mathbb{Z}^+}$ with the properties:

- $\{Y_i\}_{i \in \mathbb{Z}^+}$ is a Markov process with $Y_0 = a$ and the transition probability

$$\mathrm{P}^{(a,b)}(Y_{n+1} = dy | Y_n = x) \sim k(x,y) \frac{\nu(y)}{\lambda\, \nu(x)}\, dy \tag{2.7}$$

- $\{W_i\}_{i \in \mathbb{Z}^+}$ is the integrated Markov process with

$$W_0 = b \quad \text{and} \quad W_n = b + Y_1 + \ldots + Y_n.$$

In analogy to the Gaussian case we can state

Proposition 2.4 *For $n \in \mathbb{N}$ and $w_{-1} := b - a$, $w_0 := b$ we have*

$$\mathrm{P}^{(a,b)}\left((W_1, \ldots, W_n) \in (dw_1, \ldots, dw_n)\right) = \frac{\nu(w_n - w_{n-1})}{\lambda^n \nu(a)} e^{-\mathcal{H}_{[-1,n]}(w_{-1}, \ldots, w_n)} \prod_{i=1}^{n} dw_i \ .$$

Proof Under $\mathrm{P}^{(a,b)}$ we have already set $Y_n = W_n - W_{n-1}$, $n \geq 1$, so the law of $(W_1, \ldots, W_n)$ is determined by the law of $(Y_1, \ldots, Y_n)$. If we set $y_i := w_i - w_{i-1}$, $i \geq 2$ and $y_1 := w_1 - b$, then we have to show that under the r.h.s. of the statement the $(y_i)_{i=1,\ldots,n}$ are distributed like the first n steps of a Markov chain starting at a with the transition probability given by (2.7). The Hamiltonian can be now written in the following way

$$\mathcal{H}_{[-1,n]}(w_{-1}, \ldots, w_n) = \sum_{i=1}^{n} V_1(y_i) + V_2(y_1 - a) + \sum_{i=1}^{n-1} V_2(y_{i+1} - y_i) \ .$$

Therefore we conclude

$$\frac{\nu(w_n - w_{n-1})}{\lambda^n \nu(a)} e^{-\mathcal{H}_{[-1,n]}(w_{-1}, \ldots, w_n)} = \frac{\nu(y_n)}{\lambda^n \nu(a)}\, k(a, y_1) \prod_{i=2}^{n} k(y_{i-1}, y_i)$$

$$= \frac{\nu(y_1)}{\lambda \nu(a)}\, k(a, y_1) \prod_{i=2}^{n} \frac{\nu(y_i)}{\lambda \nu(y_{i-1})}\, k(y_{i-1}, y_i) \ ,$$

and we are done, because the last statement is just the density of the law of $(Y_1, ..., Y_n)$ under $\mathrm{P}^{(a,b)}$ w.r.t. the Lebesgue-measure $dy_1 \cdots dy_n$. □

The following density will play a crucial role later on.

Definition 2.5 *For $n \geq 2$ we define the density of (W_{n-1}, W_n) by*

$$\varphi_n^{(a,b)}(w_1, w_2) := \frac{\mathrm{P}^{(a,b)}\left((W_{n-1}, W_n) \in (dw_1, dw_2)\right)}{dw_1 dw_2}.$$

The next Proposition provides a representation for the free model, which can be seen as a bridge of the integrated Markov chain.

Proposition 2.6

$$\mathbb{P}_{0,N}(.) = \mathrm{P}^{(0,0)}((W_1, ..., W_{N-1}) \in .|W_N = W_{N+1} = 0)$$

and

$$\mathcal{Z}_{0,N} = \lambda^{N+1} \varphi_{N+1}^{(0,0)}(0,0).$$

Proof By Proposition 2.4, the r.h.s. can be written (conditional density)

$$\mathrm{P}^{(0,0)}((W_1, ..., W_{N-1}) \in .|W_N = W_{N+1} = 0)$$

$$= \frac{1}{\lambda^{N+1} \varphi_{N+1}^{(0,0)}(0,0)} \cdot \int e^{-\mathcal{H}_{[-1,N+1]}(w_{-1},...,w_{N+1})} \prod_{i=1}^{N-1} dw_i,$$

where $w_{-1} = w_0 = w_N = w_{N+1} = 0$. The first expression in the above calculation is a probability measure (conditioned on the event $\{W_N = W_{N+1} = 0\}$), so plugging in $\mathbb{R}^{N-1}$, we have

$$\iff \int_{\mathbb{R}^{N-1}} e^{-\mathcal{H}_{[-1,N+1]}(w_{-1},...,w_n)} \prod_{i=1}^{N-1} dw_i = \lambda^{N+1} \varphi_{N+1}^{(0,0)}(0,0)$$
$$\iff \mathcal{Z}_{0,N} = \lambda^{N+1} \varphi_{N+1}^{(0,0)}(0,0).$$

This concludes the proof. □

Proposition 2.7 *The right-eigenfunction ν of $\mathcal{K}$ is bounded, i.e there exists an $\kappa > 0$, such that for all $x \in \mathbb{R}$ we have $\nu(x) \leq \kappa$.*

Proof We know already that $\nu \in L^\infty(\mathbb{R}, ||.||_\infty)$, i.e. almost surely boundedness. Now by the eigenfunction equation and Hölder-inequality with $p = 1$ we obtain the result, for all $x \in \mathbb{R}$

$$\lambda \nu(x) = \int_{\mathbb{R}} k(x,y) \nu(y) \, dy \leq \int_{\mathbb{R}} k(x,y) \, dy \, ||\nu||_\infty \leq M \, ||\nu||_\infty$$

where the last step is due to the proof of Proposition 2.3. □

2.2 Markovian description of the free model

Proposition 2.8 *The Markov chain $\{Y_i\}_{i \in \mathbb{Z}^+}$ has the invariant measure $\pi(dz) = \nu(z)\, w(z)\, dz$ and it holds*

$$\int |z|\, \pi(dz) < \infty\ .$$

Proof The invariance can be seen immediately. Now as a right eigenfunction of $\mathcal{K}$ is $\nu \in L^\infty(\mathbb{R}, ||.||_\infty)$ and so $w \in L^1(\mathbb{R}, ||.||_1)$. Therefore by Hölder and condition **C4**

$$\int |x|\, w(x)\, dx = \frac{1}{\lambda} \int |x| \int e^{-V_1(x) - V_2(x-y)}\, w(y)\, dy\, dx$$
$$\leq \frac{1}{\lambda} ||w_1||_1\, ||e^{-V_2(x-\cdot)}||_\infty \int |x|\, e^{-V_1(x)}\, dx < \infty\ ,$$

then again using Hölder we arrive by

$$\int |z|\, \pi(dz) \leq ||\nu||_\infty \int |x|\, w(x)\, dx < \infty\ .$$

$\square$

Proposition 2.9
The Markov chain $\{Y_i\}_{i \in \mathbb{Z}^+}$ is uniformly bounded in the sense that there exists an $L_1 > 0$, such that for all $n \in \mathbb{N}$ we have

$$\mathbb{E}_{\mathrm{P}(0,0)} |Y_n| \leq L_1\ .$$

Proof Let $P_n(x, dy) = P_n(x, y)\, dy$ denote the n-step transition probability of $\{Y_i\}_{i \in \mathbb{Z}^+}$, starting in x and going to dy, i.e. the n-th convolution of the transition operator

$$P_1(x, dy) = \frac{e^{-V_1(y) - V_2(y-x)}\, \nu(y)}{\lambda\, \nu(x)}\ .$$

First we want to show, that for all $y \in \mathbb{R}$, $P_1(0, y) \leq c\, \nu(y)\, w(y)$ for an $c > 0$ independent of y. Thanks to continuity of the potentials we have for $y \in [-M-1, M+1]$

$$\frac{P_1(0, y)}{\nu(y)\, w(y)} \leq \frac{1}{\lambda \nu(0)} \left\| \frac{e^{-V_1(\cdot) - V_2(\cdot)}}{w(\cdot)} \mathbf{1}_{[-M-1, M+1]}(\cdot) \right\|_\infty =: \tilde{c}_1 < \infty$$

In the other case we can bound from below by

$$w(y) = \frac{1}{\lambda} \int e^{-V_1(y) - V_2(y-z)}\, w(z)\, dz$$
$$\geq \frac{e^{-V_1(y)}}{\lambda} \min \begin{cases} \int_0^1 e^{-V_2(y-z)}\, w(z)\, dz &, y > M+1 \\ \int_{-1}^0 e^{-V_2(y-z)}\, w(z)\, dz &, y < -(M+1) \end{cases}$$
$$\geq \frac{e^{-V_1(y) - V_2(y)}}{\lambda} \min \begin{cases} \int_0^1 w(z)\, dz &, y > M+1 \\ \int_{-1}^0 w(z)\, dz &, y < -(M+1) \end{cases}$$
$$=: \frac{e^{-V_1(y) - V_2(y)}}{\lambda}\, \tilde{c}$$

and therefore
$$\frac{P_1(0,y)}{\nu(y)\,w(y)} \le \frac{\tilde c}{\nu(0)} =: \tilde c_2 < \infty\,.$$
Alltogether we have with $c := \max\{\tilde c_1, \tilde c_2\}$
$$P_1(0,y) \le c\,\nu(y)\,w(y) \quad \text{for all } y \in \mathbb{R}$$
and so from Proposition 2.8 for all $n \in \mathbb{N}$ (set $P_0 := Id$)
$$\begin{aligned}\mathbb{E}_{P^{(0,0)}}|Y_n| &= \int dz\, P_n(0,z)\,|z| = \int dz \int dy\, P_1(0,y)\, P_{n-1}(y,z)\,|z|\\ &\le c \int dz \int \pi(dy)\, P_{n-1}(y,z)\,|z|\\ &= c \int \pi(dz)\,|z| < \infty\,.\end{aligned}$$
□

2.3 Bounds on the density $\varphi_N^{(0,0)}(0,0)$

Later we will see that crucial for further investigation is the behavior of the density in Definition 2.5 at $(0,0)$, when the Markov chain $\{Y_i\}_{i\in\mathbb{Z}^+}$ and the integrated Markov chain $\{W_i\}_{i\in\mathbb{Z}^+}$ start in $(a,b) = (0,0)$. The next bounds will be sufficient

Proposition 2.10 *The following lower and upper bound hold*

- *There exists a $c > 0$, such that $\frac{c}{N} \le \varphi_N^{(0,0)}(0,0)$ for all $N \in \mathbb{N}_{\ge 2}$*
- $\lim_{N\to\infty} \frac{1}{N}\log \varphi_N^{(0,0)}(0,0) \le 0$.

Observe that the lower bound is not sharp. Recalling the behavior of $\varphi_N^{(0,0)}(0,0)$ from the previous chapter, one should rather expect an order of $N^{-1/2}$. This inaccuracy is due to the rough estimate in (2.9). However, pre-empting chapter 5, even the accurate lower bound would yield at most a fourth-order phase transition, which is not enough.

Proof of the Proposition 2.10

2.3.1 The lower bound

Proposition 2.6 allows us to write
$$\varphi_N^{(0,0)}(0,0) = \frac{1}{\lambda^N}\mathcal{Z}_{0,N-1} = \frac{1}{\lambda^N} \int_{\mathbb{R}^{N-2}} e^{-\sum_{i=1}^{N} V_1(\nabla\varphi_i) - \sum_{i=0}^{N-1} V_2(\Delta\varphi_i)} \prod_{i=1}^{N-2} d\varphi_i$$

2.3 Bounds on the density $\varphi_N^{(0,0)}(0,0)$

with $\varphi_{-1} = \varphi_0 = \varphi_{N-1} = \varphi_N = 0$. For the sake of convenience we want to consider just an odd number for N. The reason is, that now we have an even number of field variables and it is possible to use a symmetry argument. In this case the boundary conditions are $\varphi_{-1} = \varphi_0 = \varphi_{2N} = \varphi_{2N+1} = 0$ and

$$\varphi_{2N+1}^{(0,0)}(0,0) = \frac{1}{\lambda^{2N+1}} \int_{\mathbb{R}^{2N-1}} e^{-\sum_{i=1}^{2N+1} V_1(\nabla \varphi_i) - \sum_{i=0}^{2N} V_2(\Delta \varphi_i)} \prod_{i=1}^{2N-1} d\varphi_i$$

Since we want to obtain a lower bound, we restrict the integration to $C_N^1(\varepsilon) := \mathbb{R}^{2N-1} \cap \{|\varphi_N - \varphi_{N-1}| < \varepsilon, |\varphi_N - \varphi_{N+1}| < \varepsilon\}$, for an arbitrary fixed $\varepsilon > 0$. On $C_N^1(\varepsilon)$ we have $|\nabla \varphi_{N+1}| < \varepsilon$ and $|\Delta \varphi_N| < 2\varepsilon$. By our assumption V_2 is continuous and so there is a $M_\varepsilon > 0$ with $V_2(x) \leq M_\varepsilon$ for all $|x| < 2\varepsilon$. Therefore we obtain

$$\varphi_{2N+1}^{(0,0)}(0,0)$$
$$\geq \frac{e^{-(V_1(0)+M_\varepsilon)}}{\lambda^{2N+1}} \int_{C_N^1(\varepsilon)} e^{-\sum_{i=1}^{N} V_1(\nabla \varphi_i) - \sum_{i=0}^{N-1} V_2(\Delta \varphi_i)} e^{-\sum_{i=N+1}^{2N} V_1(\nabla \varphi_i) - \sum_{i=N+1}^{2N} V_2(\Delta \varphi_i)} \prod_{i=1}^{2N-1} d\varphi_i$$
$$= \frac{e^{-(V_1(0)+M_\varepsilon)}}{\lambda^{2N+1}} \int_{\mathbb{R}} d\varphi_N \left[\int_{C_N^2(\varepsilon)} e^{-\sum_{i=1}^{N} V_1(\nabla \varphi_i) - \sum_{i=0}^{N-1} V_2(\Delta \varphi_i)} \prod_{i=1}^{N-1} d\varphi_i \right]^2,$$

where in the last step we have used the symmetry of V_1 and the symmetry of the integrand on $C_N^1(\varepsilon)$, setting $C_N^2(\varepsilon) := \mathbb{R}^{N-1} \cap \{|\varphi_N - \varphi_{N-1}| < \varepsilon\}$. Now we take $c_N > 0$ and make a further restriction on the integration, then we use Jensen's inequality

$$\varphi_{2N+1}^{(0,0)}(0,0)$$
$$\geq \frac{e^{-(V_1(0)+M_\varepsilon)}}{\lambda^{2N+1}} \int_{-c_N}^{c_N} d\varphi_N \left[\int_{C_N^2(\varepsilon)} e^{-\sum_{i=1}^{N} V_1(\nabla \varphi_i) - \sum_{i=0}^{N-1} V_2(\Delta \varphi_i)} \prod_{i=1}^{N-1} d\varphi_i \right]^2$$
$$\geq \frac{e^{-(V_1(0)+M_\varepsilon)}}{2c_N \lambda} \left[\frac{1}{\lambda^N} \int_{-c_N}^{c_N} d\varphi_N \int_{C_N^2(\varepsilon)} e^{-\sum_{i=1}^{N} V_1(\nabla \varphi_i) - \sum_{i=0}^{N-1} V_2(\Delta \varphi_i)} \prod_{i=1}^{N-1} d\varphi_i \right]^2$$
$$= \frac{e^{-(V_1(0)+M_\varepsilon)}}{2c_N \lambda} \left[\frac{1}{\lambda^N} \int_{-c_N}^{c_N} d\varphi_N \int_{C_N^2(\varepsilon)} \frac{\nu(0)}{\nu(\varphi_N - \varphi_{N-1})} \frac{\nu(\varphi_N - \varphi_{N-1})}{\nu(0)} \right.$$
$$\left. \cdot e^{-\sum_{i=1}^{N} V_1(\nabla \varphi_i) - \sum_{i=0}^{N-1} V_2(\Delta \varphi_i)} \prod_{i=1}^{N-1} d\varphi_i \right]^2$$
$$\geq \frac{e^{-(V_1(0)+M_\varepsilon)} (\nu(0))^2}{2c_N \lambda \kappa^2} \left[P^{(0,0)}(|W_N| \leq c_N, |W_N - W_{N-1}| \leq \varepsilon) \right]^2. \quad (2.8)$$

In the last inequality we have used Proposition 2.7 and Proposition 2.4. Now we use twice the Markov-inequality to obtain

$$P^{(0,0)}(|W_N| \leq c_N, |Y_N| \leq \varepsilon) = 1 - P^{(0,0)}(\{|W_N| > c_N\} \cup \{|Y_N| > \varepsilon\})$$
$$\geq 1 - P^{(0,0)}(|W_N| > c_N) - P^{(0,0)}(|Y_N| > \varepsilon)$$
$$\geq 1 - \frac{1}{c_N} \mathbb{E}_{P^{(0,0)}}[|W_N|] - \frac{1}{\varepsilon} \mathbb{E}_{P^{(0,0)}}|Y_N|.$$

By Proposition 2.9 we have

$$\mathbb{E}_{P^{(0,0)}}[|W_N|] \leq \sum_{i=1}^{N} \mathbb{E}_{P^{(0,0)}}|Y_N| \leq L_1 N \ . \tag{2.9}$$

Furthermore, since $c_N, \varepsilon > 0$ were arbitrary, we choose $c_N = L_2 N$ such that $L_2 > 3L_1$ and $\varepsilon = 2L_1$ to get finally

$$P^{(0,0)}(|W_N| \leq c_N \ , \ |Y_N| \leq \varepsilon) = 1 - \frac{L_1 N}{c_N} - \frac{L_1}{\varepsilon} = 1 - \frac{L_1}{L_2} - \frac{1}{2} > \frac{1}{6} \ .$$

Of course this lower bound is not optimal, but any lower bound greater than 0 suffices to conclude from (2.8) that

$$\varphi_N^{(0,0)}(0,0) \geq \frac{c}{N} \quad , \text{ where } \quad 0 < c := \frac{e^{-(V_1(0)+M_{2L_1})}(\nu(0))^2}{72 L_2 \lambda \kappa^2} \ ,$$

and this gives the desired lower bound.

2.3.2 The upper bound

The upper bound seems to be trivial because we allow even something less than exponential growth of $\varphi_N^{(0,0)}(0,0)$. Nevertheless, it requires still some work. By condition **C3** and **C4** we can "erase" $V_1(\nabla \phi_{2N+1})$, $V_1(\nabla \phi_{2N})$, $V_2(\Delta \phi_{2N})$ and $V_2(\Delta \phi_{2N-1})$ to obtain

$$\varphi_{2N+1}^{(0,0)}(0,0) = \frac{1}{\lambda^{2N+1}} \int_{\mathbb{R}^{2N-1}} e^{-\mathcal{H}_{[-1,2N+1]}(\varphi)} \prod_{i=1}^{2N-1} d\varphi_i \leq \frac{e^{-4c}}{\lambda^{2N+1}} \int_{\mathbb{R}^{2N-1}} e^{-\mathcal{H}_{[-1,2N-1]}(\varphi)} \prod_{i=1}^{2N-1} d\varphi_i$$

$$= \frac{e^{-4c}}{\lambda^2} \int_{\mathbb{R}^2} \frac{\nu(0)}{\nu(\varphi_{2N-1} - \varphi_{2N-2})} P^{(0,0)}(W_{2N-2} \in d\varphi_{2N-2}, W_{2N-1} \in d\varphi_{2N-1})$$

$$= \frac{\nu(0)}{\lambda^2} e^{-4c} \mathbb{E}_{P^{(0,0)}} \left(\frac{1}{\nu(Y_{2N-1})} \right) .$$

It only remains to prove that there exists $C > 0$, such that

$$\mathbb{E}_{P^{(0,0)}} \left(\frac{1}{\nu(Y_{2N-1})} \right) \leq C \, , \quad \forall n \in \mathbb{N} \ . \tag{2.10}$$

Now similarly to the proof of Proposition 2.9 we have

$$\mathbb{E}_{P^{(0,0)}} \left(\frac{1}{\nu(Y_n)} \right) = \int dz \, P_n(0,z) \frac{1}{\nu(z)} = \int dz \int dy \, P_1(0,y) \, P_{n-1}(y,z) \frac{1}{\nu(z)}$$

$$\leq \ldots \leq c \int dz \int \pi(dy) \, P_{n-1}(y,z) \frac{1}{\nu(z)} = c \int \pi(dz) \frac{1}{\nu(z)}$$

$$= c \int \frac{1}{\nu(z)} \nu(z) \, w(z) \, dz = c \|w\|_1 \ < \ \infty \ .$$

$\square$

2.4 The "double-contact" process

The technical tool here will be the construction of a renewal process that will help to provide a sufficient lower bound for the free energy. Fix an arbitrarily chosen $\varepsilon > 0$ and without loss of generality we take N even (for an odd N in the calculations below one should sum up to $\lfloor N/2 \rfloor$ instead of $(N-2)/2$). Using the expansion (1.22) of the product measure and restricting summation to sets $\mathfrak{A}_{2k} \subseteq \mathcal{P}(\mathbb{N}^{2k})$ defined by

$$\mathfrak{A}_{2k} := \{\{t_1 - 1, t_1, t_2 - 1, t_2, \dots, t_k - 1, t_k\} \mid 0 = t_0 < t_1 < \dots < t_k < N \text{ and } t_i - t_{i-1} \geq 2\}$$

we obtain

$$\begin{aligned}
\mathcal{Z}_{\varepsilon,N} &= \sum_{k=0}^{N-1} \varepsilon^k \sum_{\substack{A \subseteq \{1,\dots,N-1\} \\ |A|=k}} \int e^{-\mathcal{H}_{[-1,N+1]}(\varphi)} \prod_{m \in A} \delta_0(d\varphi_m) \prod_{n \in A^C} d\varphi_n \\
&\geq \sum_{k=0}^{\frac{N-2}{2}} \varepsilon^{2k} \sum_{\substack{A \subseteq \{1,\dots,N-1\} \\ |A|=2k}} \int e^{-\mathcal{H}_{[-1,N+1]}(\varphi)} \prod_{m \in A} \delta_0(d\varphi_m) \prod_{n \in A^C} d\varphi_n \\
&\geq \sum_{k=0}^{\frac{N-2}{2}} \varepsilon^{2k} \sum_{A \subseteq \mathfrak{A}_{2k}} \int e^{-\mathcal{H}_{[-1,N+1]}(\varphi)} \prod_{m \in A} \delta_0(d\varphi_m) \prod_{n \in A^C} d\varphi_n \\
&\geq \sum_{k=0}^{\frac{N-2}{2}} \varepsilon^{2k} \sum_{\substack{0=t_0<t_1<\dots<t_k<N<t_{k+1}=N+1 \\ t_i-t_{i-1}\geq 2}} \prod_{j=1}^{k+1} \widetilde{K}(t_j - t_{j-1}) \, . \quad (2.11)
\end{aligned}$$

In the last step we factorized according to $A \subseteq \mathfrak{A}_{2k}$ with

$$\widetilde{K}(n) := \begin{cases} e^{-\mathcal{H}_{[-1,2]}(0,0,0,0)} = e^{-2(V_1(0)+V_2(0))} & , n = 2 \\ \int_{\mathbb{R}^{n-2}} e^{-\mathcal{H}_{[-1,n]}(w_{-1},\dots,w_n)} dw_1 \cdots dw_{n-2} & , n \geq 3 \\ \text{with } w_{-1} = 0, w_0 = 0, w_{n-1} = 0, w_n = 0 \, . \end{cases}$$

We can now choose a $\mu_\varepsilon > 0$ and for $n \in \mathbb{N}_{\geq 2}$ set

$$K^{\mu_\varepsilon}(n) := \frac{\varepsilon^2}{\lambda^n} \widetilde{K}(n) e^{-\mu_\varepsilon n} \left(= \varepsilon^2 \varphi_n^{(0,0)}(0,0) e^{-\mu_\varepsilon n} \right)$$

such that

$$\sum_{n \in \mathbb{N}_{\geq 2}} K^{\mu_\varepsilon}(n) = 1 \, .$$

Indeed, due to the lower bound in Proposition 2.10 we have $\mu_\varepsilon > 0$ for all $\varepsilon > 0$. Now we define a probability measure $\mathcal{P}^{\mu_\varepsilon}$ and a renewal process $\{\eta_i\}_{i \in \mathbb{Z}^+}$ on $\mathbb{N}_{\geq 2}$. More precisely a process starting in $\eta_0 = 0$ and having i.i.d. increments $\{\eta_{k+1} - \eta_k\}_k$ and the inter-arrival law

$$\mathcal{P}^{\mu_\varepsilon}(\eta_{k+1} - \eta_k = n) = K^{\mu_\varepsilon}(n) \, .$$

We stress also that this process can be written as

$$\eta_0 = 0, \qquad \eta_{k+1} = \inf\{i \geq \eta_k + 2 \mid \varphi_{i-1} = \varphi_i = 0\}.$$

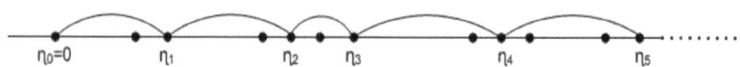

Figure 2.1: A sketch of the double contacts $\{\eta_i\}_{i\in\mathbb{Z}^+}$. We have imposed $\eta_{k+1} - \eta_k \geq 2$.

Observe that by independence

$$\prod_{j=1}^{k+1} K^{\mu_\varepsilon}(t_j - t_{j-1}) = \mathcal{P}^{\mu_\varepsilon}(\eta_1 = t_1, \eta_2 - \eta_1 = t_2 - t_1, ..., \eta_{k+1} - \eta_k = N + 1 - t_k)$$

$$= \mathcal{P}^{\mu_\varepsilon}(\eta_1 = t_1, \eta_2 = t_2, ..., \eta_{k+1} = N + 1)$$

and therefore from (2.11) and recalling that $\eta_{k+1} - \eta_k \geq 2$ we obtain

$$\mathcal{Z}_{\varepsilon,N} \geq \sum_{k=0}^{\frac{N-2}{2}} \frac{\lambda^{N+1} e^{\mu_\varepsilon (N+1)}}{\varepsilon^2} \sum_{\substack{0=t_0<t_1<...<t_k<N<t_{k+1}=N+1 \\ t_i-t_{i-1}\geq 2}} \mathcal{P}^{\mu_\varepsilon}(\eta_1 = t_1, \eta_2 = t_2, ..., \eta_{k+1} = N + 1)$$

$$= \frac{\lambda^{N+1} e^{\mu_\varepsilon (N+1)}}{\varepsilon^2} \sum_{k=0}^{\frac{N-2}{2}} \mathcal{P}^{\mu_\varepsilon}(\eta_{k+1} = N + 1)$$

$$= \frac{\lambda^{N+1} e^{\mu_\varepsilon (N+1)}}{\varepsilon^2} \sum_{k=1}^{\infty} \mathcal{P}^{\mu_\varepsilon}(\eta_k = N + 1)$$

$$= \frac{\lambda^{N+1} e^{\mu_\varepsilon (N+1)}}{\varepsilon^2} \mathcal{P}^{\mu_\varepsilon}(N + 1 \in \eta). \tag{2.12}$$

In the last but one equation we substituted $N/2$ by ∞ since from here on the sum is empty, due to $\eta_{1+N/2} \geq N + 2$ ($\eta_1 \geq 2, \eta_2 \geq 4$ and so on). Furthermore η denotes the set of restricted double-contacts:

$$\eta := \{k \in \mathbb{Z}^+ \mid \varphi_{k-1} = \varphi_k = 0\} \cap \mathfrak{A}_{2k}.$$

2.5 Lower bound for the free energy

First of all we recall the definition of free energy in our setting

$$F(\varepsilon) = \lim_{N\to\infty} \frac{1}{N} \log \frac{\mathcal{Z}_{\varepsilon,N}}{\mathcal{Z}_{0,N}}.$$

2.5 LOWER BOUND FOR THE FREE ENERGY

It is well defined for all $\varepsilon > 0$, as can be shown again by a standard super-additivity argument, cf. chapter 0. Observe that by Proposition 2.6 and both bounds in Proposition 2.10 it holds

$$\lim_{N\to\infty} \frac{1}{N} \log \mathcal{Z}_{0,N} = \lim_{N\to\infty} \frac{1}{N} \left((N+1)\log \lambda + \log \varphi_{N+1}^{(0,0)}(0,0) \right) = \log \lambda \ .$$

Therefore for an arbitrarily chosen $\varepsilon > 0$, by (2.12) we obtain

$$\lim_{N\to\infty} \frac{1}{N} \log \frac{\mathcal{Z}_{\varepsilon,N}}{\mathcal{Z}_{0,N}} \geq \lim_{N\to\infty} \frac{1}{N} \log \left[\frac{\lambda^{N+1} e^{\mu_\varepsilon (N+1)}}{\varepsilon^2} \mathcal{P}^{\mu_\varepsilon}(N+1 \in \eta) \right] - \lim_{N\to\infty} \frac{1}{N} \log \mathcal{Z}_{0,N}$$
$$\geq \mu_\varepsilon + \lim_{N\to\infty} \frac{1}{N} \log \left[\mathcal{P}^{\mu_\varepsilon}(N+1 \in \eta) \right] \qquad (2.13)$$

As we have already mentioned above, for all $\varepsilon > 0$ we have $\mu_\varepsilon > 0$. So, if we can show that

$$\lim_{N\to\infty} \frac{1}{N} \log \left[\mathcal{P}^{\mu_\varepsilon}(N+1 \in \eta) \right] \geq 0$$

then we have proven the trivial phase transition. Now, since $\mathcal{P}^{\mu_\varepsilon}(\eta_1 = n) = K^{\mu_\varepsilon}(n) > 0$ for all $n \in \mathbb{N}_{n\geq 2}$ the process $\{\eta_k\}_{k\in\mathbb{Z}^+}$ is aperiodic and by the Classical Renewal Theorem of [3] (chapter I, Thm. 2.2)

$$\mathcal{P}^{\mu_\varepsilon}(N+1 \in \eta) \xrightarrow[N\to\infty]{} \frac{1}{m_\varepsilon} \ ,$$

where $m_\varepsilon = \sum_{n\geq 2} n\, K^{\mu_\varepsilon}(n)$. Furthermore it is by Proposition 2.10

$$0 < m_\varepsilon = \varepsilon^2 \sum_{n\geq 2} n\, \varphi_n^{(0,0)}(0,0)\, e^{-\mu_\varepsilon n} < \infty \ .$$

Therfore as soon as $\varepsilon > 0$ the free energy is also strictly positive. This means we have proven a trivial phase transition for our pinning-model with general potentials.

3 The polymer above a "hard wall" : entropic repulsion in Gaussian case

3.1 The model with a wall

In this chapter we study a further extension of the pinning model (1.1), which now will additionally interact with a neutral hard wall. This interaction is also known as the phenomenon of entropic repulsion. The presence of the hard wall in $\{1, ..., N\text{-}1\}$ is modeled by the positivity constraint $\widetilde{\Omega}_N^+ := \{\varphi_i \geq 0 \,|\, i \in \{1, ..., N-1\}\}$. The measure describing this process is then the conditional measure $\mathbb{P}_{\varepsilon,N}^w$

$$\mathbb{P}_{\varepsilon,N}^w(\cdot) := \mathbb{P}_{\varepsilon,N}\left(\cdot \,|\, \widetilde{\Omega}_N^+\right) , \qquad (3.1)$$

where $\mathbb{P}_{\varepsilon,N}$ denotes the "pure"-pining measure in (1.1).

Now putting it in a formal way, (3.1) is described by the distribution on $\mathbb{R}^{N-1}$:

$$\mathbb{P}_{\varepsilon,N}^w(d\varphi) := \frac{\exp(-\mathcal{H}_{[-1,N+1]}(\varphi))}{\mathcal{Z}_{\varepsilon,N}^w} \prod_{i=1}^{N-1} \left(\varepsilon\delta_0(d\varphi_i) + d\varphi_i \,\mathbb{1}_{\{\varphi_i \geq 0\}}\right) , \qquad (3.2)$$

where the Hamiltonian is of the following form ($\alpha, \beta > 0$)

$$\mathcal{H}_{[-1,N+1]}(\varphi_{-1}, ..., \varphi_{N+1}) = \frac{\alpha}{2}\sum_{i=1}^{N+1}(\nabla\varphi_i)^2 + \frac{\beta}{2}\sum_{i=0}^{N}(\Delta\varphi_i)^2$$

and for simplicity we impose again zero boundary conditions, i.e.

$$\varphi_{-1} = \varphi_0 = \varphi_N = \varphi_{N+1} = 0 .$$

64   THE POLYMER ABOVE A "HARD WALL" : ENTROPIC REPULSION IN GAUSSIAN CASE

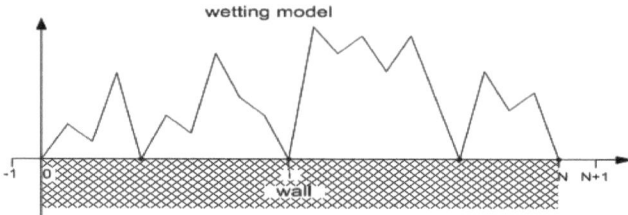

Figure 3.1: A sketch of the wetting model $\mathbb{P}^w_{\varepsilon,N}$. This time the polymer-chain fluctuates above an impenetrable wall (membrane). On the one hand it is attracted by the intergersites at the x-axis, on the other hand it is repelled by the wall through its own fluctuations.

The additional effect of an impermeable wall, which the chain is not allowed to cross, is known in the context of interface models, confer for instance [29]. This model is then usually called the wetting model, since by means of localization one is interested if the wet-phase between the wall and the interface (here the chain) prevails or not. By the definition (3.2) one can immediately see that indeed :

$$\mathbb{P}^w_{\varepsilon,N}(d\varphi_1 \cdots d\varphi_{N-1}) = \mathbb{P}_{\varepsilon,N}\left(d\varphi_1 \cdots d\varphi_{N-1} \mid \varphi_1 \geq 0, \ldots, \varphi_{N-1} \geq 0\right) \ .$$

3.2   The main result on wetting

We remark that similarly to the pinning, the wetting model undergoes two opposite effects, the entropy and the energy. However, this time those effects are represented on the one hand by the self-interaction and the wall and on the other hand by pinning towards the defect-line. The additional effect of the wall is of repulsive nature, since the polymer can achieve space for its fluctuations just by repelling itself away from the wall. Especially in view of the previous results in the pinning case it is therefore interesting, if the wall-constraint is strong enough to obtain this time a (proper) phase transition.

We refer to the previous chapters for definitions of localization/delocalization. Recall the free energy in the setting of the wetting model

$$F^w(\varepsilon) = \lim_{N \to \infty} \frac{1}{N} \log \left(\frac{\mathcal{Z}^w_{\varepsilon,N}}{\mathcal{Z}^w_{0,N}} \right) \quad , \ \varepsilon \geq 0 \ .$$

So far we had to realize that fluctuations of our model are not strong enough, meaning that an arbitrarily small pinning strength localizes the behavior of the chain. This happens even when, comparing to the ∇-potential, we impose very strong Laplacian-potentials in the model. This behavior changes finally dramatically, if the chain meets an impermeable wall.

3.3 The free wetting model

Theorem 3.1 (Localization in wetting case) *For every $\alpha, \beta > 0$ there exists $\varepsilon_c^+ > 0$, such that the model $\mathbb{P}_{\varepsilon,N}^w$ reveals the following delocalization-localization behavior:*

$$\mathcal{D} = [0, \varepsilon_c^+] \quad \text{and} \quad \mathcal{L} = (\varepsilon_c^+, \infty) \ .$$

Furthermore, on $\mathcal{L}$ the free energy is real analytic and

$$F^w(\varepsilon) \sim \log \varepsilon \quad , \varepsilon \to \infty \ .$$

The gradient model ($\beta = 0$) has been studied for instance by [7] for Gaussian and by [12] for general potentials, cf. also [1], [13], [16], [18], [20]. The Laplacian case ($\alpha = 0$) was investigated for general potentials in [10]. In both cases the critical point turns out to be strictly positive, i.e. the models exhibit also a non-trivial phase transition. At this point we remark again that [10] was the most important paper for the methods developed here.

3.3 The free wetting model

First of all we would like to describe the free model, i.e. $\mathbb{P}_{0,N}^w$. There is a connection to the same Markov chain, which was constructed in the subsection 1.4.1 of the pinning model.

Let us define the orthant set of the integrated Markov chain $\{W_i\}_{i \in \mathbb{Z}^+}$ defined in the subsection 1.4.1

$$\Omega_N^+ := \{W_1 \geq 0, ..., W_N \geq 0\} \ . \tag{3.3}$$

Similarly to the density in pinning case we introduce

Definition 3.2 *For $n \geq 2$ we define the conditional density of (W_{n-1}, W_n) by*

$$\overset{+}{\varphi}_n^{(a,b)}(w_1, w_2) := \frac{\mathrm{P}^{(a,b)}\left((W_{n-1}, W_n) \in (dw_1, dw_2) \mid \Omega_n^+\right)}{dw_1 dw_2} \ .$$

This density enables us to write our free model $\mathbb{P}_{0,N}^w$ in the following way:

Proposition 3.3

$$\mathbb{P}_{0,N}^w = \mathrm{P}^{(0,0)}\left((W_1, ..., W_{N-1}) \in . | \Omega_{N-1}^+, W_N = W_{N+1} = 0\right)$$

and

$$\mathcal{Z}_{0,N}^w = \lambda^{N+1} \overset{+}{\varphi}_{N+1}^{(0,0)}(0,0) \, \mathrm{P}^{(0,0)}(\Omega_{N-1}^+) \ .$$

Proof We can write (conditional density)

$$\mathrm{P}^{(0,0)}((W_1, ..., W_{N-1}) \in . | \Omega_{N-1}^+, W_N = W_{N+1} = 0)$$

$$= \frac{1}{\lambda^{N+1} \overset{+}{\varphi}_{N+1}^{(0,0)}(0,0)} \cdot \int_{.\cap \Omega_{N-1}^+} e^{-\mathcal{H}_{[-1,N+1]}(w_{-1},...,w_{N+1})} \prod_{i=1}^{N-1} dw_i,$$

where $w_{-1} = w_0 = w_N = w_{N+1} = 0$. The first expression in the above calculation is a probability measure (conditioned on the event $\{\Omega_{N-1}^+, W_N = W_{N+1} = 0\}$), so plugging in Ω_{N-1}^+, we obtain

$$\iff \frac{\int_{\Omega_{N-1}^+} e^{-\mathcal{H}_{[-1,N+1]}(w_{-1},\ldots,w_n)} \prod_{i=1}^{N-1} dw_i}{\mathcal{Z}_{0,N}^w} = \lambda^{N+1} \overset{+(0,0)}{\varphi}_{N+1}(0,0) \, \mathrm{P}^{(0,0)}(\Omega_{N-1}^+)$$
$$= \lambda^{N+1} \overset{+(0,0)}{\varphi}_{N+1}(0,0) \, \mathrm{P}^{(0,0)}(\Omega_{N-1}^+) \,.$$

This concludes the proof. □

The last Proposition makes it possible to represent our free model as a "bridge" of the process $\{W_i\}_{i \in \mathbb{Z}^+}$, conditioned to stay positive.

3.3.1 Convenient representation for the integrated Markov chain

We have seen in the pinning case that the density in (1.21) has played a central role. Since we are dealing with the Gaussian case it is clear that covariances of the process $\{W_k\}_{k \in \mathbb{Z}^+}$ are of special interest in order to obtain a representation for this density. Contrarily to the laborious method of inverting a band matrix in chapter 1, we can receive a more convenient representation in a more direct and easier way. Recall that we assume always $\alpha, \beta > 0$, if not explicitly stated.

In the Gaussian case we have found a very nice representation of our Markov chain $\{Y_i\}_{i \in \mathbb{Z}^+}$ and the integrated Markov chain $\{W_i\}_{i \in \mathbb{Z}^+}$, defined in the subsection 1.4.1. More precisely, if we start in $(Y_0, W_0) = (a, b)$, then under $\mathrm{P}^{(a,b)}$

$$Y_n = \gamma^n a + \gamma^{n-1} \varepsilon_1 + \ldots + \gamma^0 \varepsilon_n \tag{3.4}$$

and

$$W_n = \sum_{i=1}^n Y_i = \gamma \, r_{n-1} a + b + r_{n-1} \varepsilon_1 + \ldots + r_0 \varepsilon_n \tag{3.5}$$

where (sgn denotes the signum function):

$$r_{n-i} := \sum_{i=0}^{n-i} \gamma^i \quad \text{and} \quad \gamma = \left(\frac{\alpha + 2\beta - \sqrt{\alpha}\sqrt{\alpha + 4\beta}}{\alpha + 2\beta + \sqrt{\alpha}\sqrt{\alpha + 4\beta}} \right)^{1/2} \cdot \mathrm{sgn}(\beta) \,.$$

The process $\{\varepsilon_i\}_{i \in \mathbb{Z}^+}$ is an i.i.d. sequence of centered Gaussian variables $\sim \mathcal{N}(0, \sigma^2)$, $\sigma^2 = 1/\sigma_+$ with σ_+ defined in Proposition 1.5.

Remark 3.4
The values for γ are dependent on α and β and we will distinguish between three cases:

(i) $0 < \gamma < 1$, for the mixed model ($\alpha, \beta > 0$)
(ii) $\gamma = 0$, in ∇-case ($\beta = 0$)
(iii) $\gamma = 1$, in Δ-case ($\alpha = 0$).

Considering (3.5) in the last two cases $\{W_k\}_{k\in\mathbb{Z}^+}$ represents either a random walk or an integrated random walk. Confer Remark 1.4 and Remark A.6 for $\beta < 0$.

In case of $0 \leq \gamma < 1$ we can write the integrated Markov (under $P^{(a,b)}$) chain as

$$W_n = \gamma \frac{1-\gamma^n}{1-\gamma} a + b + \frac{1}{1-\gamma}(\varepsilon_1 + \ldots + \varepsilon_n) - \frac{1}{1-\gamma}(\gamma^n \varepsilon_1 + \gamma^{n-1}\varepsilon_2 + \ldots + \gamma\varepsilon_n) \ .$$

This representation seems to be surprising at a first glance and we were very happy to have found it. The reason is that it will be very convenient for further analysis. For the justification of the upper representation confer Appendix A.4.

3.4 Entropic repulsion

This section is devoted to the problem of entropic repulsion, which results as an effect of introducing an impermeable wall in our model. In this context let us define an important term $w_{x,y}(N)$ that we call the "conditional orthant probability".

Definition 3.5 *For $N \geq 3$ and $x, y \in \mathbb{R}$ we define the quantity*

$$w_{x,y}(N) := P^{(-x,0)}(W_n \geq 0, \, n = 1, \ldots, N-2 \mid W_{N-1} = y, \, W_N = 0)\mathbb{1}_{\{x,y\geq 0\}} \ .$$

As a remark we mention that

$$\varphi_{N+1}^{+(0,0)}(0,0) = \frac{1}{\lambda^{N+1} P^{(0,0)}(\Omega_{N-1}^+)} \frac{\mathcal{Z}_{0,N}^w}{\mathcal{Z}_{0,N}} \ \mathcal{Z}_{0,N} = \varphi_{N+1}^{(0,0)}(0,0) \, w_{0,0}(N+1) \frac{1}{P^{(0,0)}(\Omega_{N-1}^+)} \ ,$$

with the density being already defined in the pure-pinning model:

$$\varphi_n^{(a,b)}(w_1, w_2) = \frac{P^{(a,b)}((W_{n-1}, W_n) \in (dw_1, dw_2))}{dw_1 dw_2} \quad , \ n \geq 2 \ .$$

We will see later that the quantity $w_{\cdot,\cdot}(\cdot)$ plays an important role in the investigation of the wetting model. For our purposes it will be necessary to control $w_{x,y}(n)$ in all variables $x, y \in \mathbb{R}_+$ and $n \in \mathbb{N}_{\geq 2}$. This non-trivial problem is known as entropic repulsion in physical, but also in mathematical literature, e.g. [29] in the context of random interfaces. For the $(d+1)$ gradient model we refer to [6] for $d \geq 3$ and to [12] for $d = 1$. For the $(d+1)$-Laplacian case with Gaussian potentials confer [27],[22] for $d \geq 5$ and [23] for $d = 4$. We remark that in lower dimensions the decay in N for the entropic repulsion takes place on different (polynomial) scales, cf. [10] for the Laplacian model with general potentials and $d = 1$. In the following we will treat the entropic repulsion for our mixed case. We will give sufficient lower and upper bounds for $w_{x,y}(n)$ in the next subsections.
Before attempting that, let us first see what can be said in the much more easier non-conditional case. For this purpose we refer to the representation (3.5) for W_n and set $\mathbb{P} := P^{(0,0)}$. We define $S_0^\nabla = S_0^\Delta := 0$ and

$$S_n^\nabla = \varepsilon_1 + \cdots + \varepsilon_n \quad and \quad S_n^\Delta = n\varepsilon_1 + (n-1)\varepsilon_2 + \cdots + \varepsilon_n \ .$$

We have the following inequalities for the entropic repulsion

Proposition 3.6 *For all $n \in \mathbb{N}$ it holds*
$$\mathbb{P}(S_1^\nabla \geq 0, ..., S_n^\nabla \geq 0) \leq \mathbb{P}(W_1 \geq 0, ..., W_n \geq 0) \leq \mathbb{P}(S_1^\Delta \geq 0, ..., S_n^\Delta \geq 0) \ .$$

Proof Let us consider first the left-hand-side inequality. We will show by induction that
$$\bigcap_{i=1}^{n} \left\{ S_i^\nabla \geq 0 \right\} \subseteq \bigcap_{i=1}^{n} \{W_i \geq 0\} \ .$$
For $n = 1$ it is trivial, since $S_1^\nabla \geq 0$ includes $W_1 \geq 0$. Now observe that
$$W_{n+1} = (1 + \gamma + \cdots + \gamma^n)\varepsilon_1 + \cdots + \varepsilon_{n+1} = S_{n+1}^\nabla + \gamma W_n$$
and therefore by induction
$$\bigcap_{i=1}^{n+1} \left\{ S_i^\nabla \geq 0 \right\} = \bigcap_{i=1}^{n} \left\{ S_i^\nabla \geq 0 \right\} \cap \{W_{n+1} - \gamma W_n \geq 0\} \subseteq \bigcap_{i=1}^{n} \{W_i \geq 0\} \cap \{W_{n+1} - \gamma W_n \geq 0\}$$
$$\subseteq \bigcap_{i=1}^{n} \{W_i \geq 0\} \cap \{W_{n+1} \geq 0\} \subseteq \bigcap_{i=1}^{n} \{W_i \geq 0\} \ .$$

Now for the second inequality in the Proposition again $n = 1$ is trivial. Here we observe that for all $n \in \mathbb{N}$
$$S_n^\Delta = W_n + (1 - \gamma)(W_1 + \cdots + W_{n-1}) \ ,$$
since
$$S_n^\Delta = n\varepsilon_1 + (n-1)\varepsilon_2 + \cdots + \varepsilon_n = (1 + \gamma + \cdots + \gamma^{n-1})\varepsilon_1 + (1 + \gamma + \cdots + \gamma^{n-2})\varepsilon_2 + \cdots + \varepsilon_n$$
$$+ \varepsilon_1(1 - \gamma + 1 - \gamma^2 + \cdots + 1 - \gamma^{n-1}) + \varepsilon_2(1 - \gamma + 1 - \gamma^2 + \cdots + 1 - \gamma^{n-2}) + \cdots + \varepsilon_{n-1}(1 - \gamma)$$
$$= W_n + (1 - \gamma)\left[\varepsilon_1\left(\frac{1-\gamma}{1-\gamma} + \frac{1-\gamma^2}{1-\gamma} + \cdots + \frac{1-\gamma^{n-1}}{1-\gamma}\right) + \varepsilon_2\left(\frac{1-\gamma}{1-\gamma} + \frac{1-\gamma^2}{1-\gamma} + \cdots + \frac{1-\gamma^{n-2}}{1-\gamma}\right)\right.$$
$$\left. + \cdots + \varepsilon_{n-1}\right]$$
$$= W_n + (1 - \gamma)(W_1 + \cdots + W_{n-1}) \ .$$

This yields
$$\bigcap_{i=1}^{n+1} \{W_i \geq 0\} = \bigcap_{i=1}^{n} \{W_i \geq 0\} \cap \{W_{n+1} \geq 0\}$$
$$= \bigcap_{i=1}^{n} \{W_i \geq 0\} \cap \left\{ S_{n+1}^\Delta - (1-\gamma)(W_n + \cdots + W_1) \geq 0 \right\}$$
$$\subseteq \bigcap_{i=1}^{n} \{W_i \geq 0\} \cap \left\{ S_{n+1}^\Delta \geq 0 \right\} \subseteq \bigcap_{i=1}^{n+1} \left\{ S_i^\Delta \geq 0 \right\} \ .$$
$\square$

This means in particular that for Ω_N^+ in definition 3.3 there exist two constants $k_1, k_2 > 0$ such that for all $N \in \mathbb{N}$
$$\frac{k_1}{N^{1/2}} \leq \mathbb{P}(\Omega_N^+) \leq \frac{k_2}{N^a} \ ,$$

3.4 ENTROPIC REPULSION

for some $0 < a \leq 1/4$, cf. [10]. In this case the upper inequality of course cannot be sharp. After this short result in the non-conditional case, we will face the difficult part of conditional entropic repulsion.

3.4.1 Upper bound

The upper bound for $w_{x,y}(n)$ is more delicate than the lower bound and we have spent quite a lot of time on how to prove it. The reason is that not only the dependence on x, y is important, but also a "nearly" sharp bound is needed. The main idea of the proof is based on a certain "decoupling"- argument, which allows us to "decouple" a random walk part from our intergrated Markov chain. We have then to control some rest terms in a suitable manner.

What makes life difficult is conditioning to come back to some points. This causes of course some additional problems in treating the entropic repulsion. Luckily we are in a Gaussian setting and the idea here is to obtain a representation for the conditional process

$$(W_n | W_N = y, W_{N+1} = 0) \ .$$

More precisely, we have to construct $\{\widehat{W}_n(y, N)\}_{n=1,\ldots,N-1}$ in such a way, that also the whole path is taken into account, i.e. for every $N \geq 2$

$$\mathscr{L}\left((\widehat{W}_1(y, N), \ldots, \widehat{W}_{N-1}(y, N))\right) = \mathscr{L}\left(((W_1, \ldots, W_{N-1}) \,|\, W_N = y, W_{N+1} = 0)\right) \quad (3.6)$$

If we would have such a process, we could immediately state that

$$P^{(-x,0)}\left(\Omega_{N-1}^+ \,|\, W_N = y, W_{N+1} = 0\right) = P^{(-x,0)}\left(\widehat{\Omega}_{N-1}^+(y)\right) \ , \quad (3.7)$$

where $\widehat{\Omega}_n^+(y) := \{\widehat{W}_i(y, n+1) \geq 0, \, i = 1, \ldots, n\}$. This will be very helpful later on. Indeed a lot of computations lead us to

Proposition 3.7 *Under* $P^{(-x,0)}$ *the conditional integrated Markov chain that satisfies* (3.6) *has the representation*

$$\widehat{W}_n(y, N) = W_n - (W_N - y)r_1(n) - W_{N+1}r_2(n) \ , \ n = 1, \ldots, N-1 \quad (3.8)$$

where

$$r_1(n) = \frac{s_1(n)}{r(n)} \quad \text{and} \quad r_2(n) = \frac{s_2(n)}{r(n)} \ .$$

For explicit representation of $r_1(n), r_2(n)$ *confer Appendix A.5. The terms* $r_1(n), r_2(n)$ *depend of course also on* N *and* γ, *but for readable reasons we have shortened the notation. Also for* (3.8) *we write sometimes simply* $\widehat{W}_n$, *when* y *and* N *are fixed.*

Now let us define under $P^{(0,0)}$ the random walk

$$S_n := \sum_{k=1}^n \tilde{\varepsilon}_k \quad (3.9)$$

starting in $S_0 := 0$, where $\{\tilde{\varepsilon}_i\}_{i \in \mathbb{Z}^+}$ is i.i.d. Gaussian $\sim \mathcal{N}(0, \sigma^2/(1-\gamma)^2)$, cf. also subsection 3.3.1. In the next Proposition we give a conditional process $\{\widehat{S}_n(y, N)\}_{n=1,...,N-1}$ that fulfills

$$\mathscr{L}\left((\widehat{S}_1(y,N), ..., \widehat{S}_{N-1}(y,N))\right) = \mathscr{L}\left(((S_1, ..., S_{N-1}) \,|\, S_N = y)\right) . \qquad (3.10)$$

Proposition 3.8 *The conditional random walk $\widehat{S}_n(y, N)$ satisfying (3.10), has the representation*

$$\widehat{S}_n(y, N) = S_n - \frac{n}{N}(S_N - y) , \quad n = 1, ..., N-1 .$$

We write again simply $\widehat{S}_n$, when y and N are fixed.

Remark 3.9
The proof of both Propositions above is based on the representation of the conditional distributions of $(Y_n | W_N = y, W_{N+1} = 0)$ and $(S_n | S_N = y)$, respectively. These are Gaussian processes, therefore with some effort one can obtain their conditional means and covariances. Then one has to verify that those means and covariances indeed equal those ones of the direct representations in Proposition 3.7 and Proposition 3.8. We defer the proof to Appendix A.5.

Next it will be useful to get rid of the dependence on $x \geq 0$.

Lemma 3.10 *There exists an $N_0 \in \mathbb{N}$, such that for all $N \in \mathbb{N}_{\geq N_0}$ the "conditional orthant probability" in Definition 3.5 attains its maximum at $x = 0$, i.e.*

$$w_{x,y}(N+1) \leq w_{0,y}(N+1) \quad , \; y \in \mathbb{R}_+ .$$

Proof The idea is to show that there exists $q(\gamma, n, N) > 0$ such that

$$\left\{\widehat{W}_n(y)\right\}_{n=1,...,N-1} \text{ under } \mathrm{P}^{(-x,0)} \stackrel{d}{=} \left\{\widehat{W}_n(y) - q(\gamma, n, N)x\right\}_{n=1,...,N-1} \text{ under } \mathrm{P}^{(0,0)} . \qquad (3.11)$$

We defer this part to Appendix A.5 and conclude that for $N \geq N_0$ and $x, y \geq 0$

$$\begin{aligned} w_{x,y}(N+1) &= \mathrm{P}^{(-x,0)}\left(\widehat{\Omega}^+_{N-1}(y)\right) \\ &= \mathrm{P}^{(0,0)}\left(\widehat{W}_1(y) \geq q(\gamma, 1, N)x, ..., \widehat{W}_{N-1}(y) \geq q(\gamma, N-1, N)x\right) \\ &\leq w_{0,y}(N+1) . \end{aligned}$$

$\square$

Remark 3.11 *As a consequence of the last Lemma, from now on we will concentrate on the case $x = 0$ and denote*

$$\mathbb{P} := \mathrm{P}^{(0,0)} .$$

3.4 ENTROPIC REPULSION

The problem we have now is how to treat the conditional integrated Markov chain. In order to handle this, it took some effort to rewrite it in a specific way. As a first step we obtain from (3.8) and (3.5) the representation

$$\widehat{W}_n = \widetilde{S}_n - \widehat{U}_n \,,$$

where

$$\widetilde{S}_n := \widehat{S}_n + y\left(r_1(n) - \frac{n}{N}\right) \,, \quad S_n \sim \mathcal{N}\left(0, n\frac{\sigma^2}{(1-\gamma)^2}\right)$$

$$\widehat{U}_n := \widetilde{S}_n - \widehat{W}_n = S_N\left(-\frac{n}{N} + r_1(n) + r_2(n)\right) + U_n - U_N r_1(n) - U_{N+1} r_2(n)$$

$$U_n := \frac{1}{1-\gamma}(\gamma^n \varepsilon_1 + \gamma^{n-1}\varepsilon_2 + \dots + \gamma \varepsilon_n) \tag{3.12}$$

and $\widehat{S}_n$ is our conditional RW defined in Proposition 3.8.

Having rewritten the conditional process $\widehat{W}$, we use now the following "decoupling"-argument. For a positive and strictly increasing sequence f_N we can estimate from above

$$\mathbb{P}\left(\widehat{W}_1 \geq 0, ..., \widehat{W}_{N-1} \geq 0\right) \leq \mathbb{P}\left(\inf_{n=1,...,N-1} \widetilde{S}_n \geq -f_N\right) + \mathbb{P}\left(\sup_{n=1,...,N-1} |\widehat{U}_n| \geq f_N\right) \tag{3.13}$$

We will see that choosing a specific sequence f_N will give us sufficient upper bounds on both expressions on the right hand side. We start with the first one by considering the

Lemma 3.12 *For every $c > 0$ there exists an $K_c \geq 0$ such that for all $N \geq 2$*

$$\mathbb{P}(S_k \geq -c, \forall 1 \leq k \leq N-1 \mid S_N = 0) \leq \frac{K_c}{N} \,. \tag{3.14}$$

Proof Although $\{S_n\}_{n \geq 0}$ doesn't have variance one, cf. (3.12), wlog we can assume for the proof that $\{S_n\}_{n \geq 0}$ is a standard Gaussian random walk. Let $c > 0$, we know that

$$\mathbb{P}(S_k \geq 0, \forall 1 \leq k \leq N-1 \mid S_N = 0) = \frac{1}{N} \,,$$

confer for instance [13]. Denote by $\mathbb{P}_x$ the law of the random walk started at x. Recall that under $\mathbb{P}_x$ the value of the density of S_n at y is

$$\frac{1}{\sqrt{2\pi n}} e^{-\frac{(y-x)^2}{2n}} \,.$$

With that let us consider the following lower bound:

$$\frac{1}{N+2} = \mathbb{P}(S_k \geq 0, \forall 1 \leq k \leq N+1 \mid S_{N+2} = 0)$$

$$\geq \mathbb{P}(S_1, S_{N+1} \in [c, c+1],\ S_k \geq 0, \forall 2 \leq k \leq N \mid S_{N+2} = 0)$$

$$= \sqrt{2\pi(N+2)} \int_c^{c+1} \int_c^{c+1} \frac{e^{-x^2/2}}{\sqrt{2\pi}} \mathbb{P}_x(S_k \geq 0, \forall 1 \leq k \leq N-1 \mid S_N = y) \tag{3.15}$$

$$\cdot \frac{e^{-(y-x)^2/(2N)}}{\sqrt{2\pi N}} \frac{e^{-y^2/2}}{\sqrt{2\pi}} \, dx \, dy \,,$$

where in the last step we have used the Markov property. Now observe, that the process $\{S_k\}_{0\leq k\leq N}$ under the law $\mathbb{P}_x(\,\cdot\,|S_N=y)$ has the same law as the process $\{x+S_k+\frac{k}{N}(y-x)\}_{0\leq k\leq N}$ under the law $\mathbb{P}(\,\cdot\,|S_N=0)$, therefore

$$\mathbb{P}_x(S_k \geq 0,\, \forall 1 \leq k \leq N-1 | S_N = y) = \mathbb{P}(S_k \geq -x - \tfrac{k}{N}(y-x),\, \forall 1 \leq k \leq N-1 \mid S_N = 0).$$

For $x, y \geq c$ we have

$$x + \frac{k}{N}(y-x) = \left(1 - \frac{k}{N}\right)x + \frac{k}{N}y \geq c,$$

hence

$$\mathbb{P}\left(S_k \geq -x - \frac{k}{N}(y-x),\, \forall 1 \leq k \leq N-1 \mid S_N = 0\right) \geq \mathbb{P}\left(S_k \geq -c,\, \forall 1 \leq k \leq N-1 \mid S_N = 0\right).$$

Coming back to (3.15), we can write

$$\frac{1}{N+2} \geq \frac{1}{2\pi}\sqrt{\frac{N+2}{N}}\, e^{-(c+1)^2 - \frac{1}{2N}}\, \mathbb{P}(S_k \geq -c,\, \forall 1 \leq k \leq N-1 \mid S_N = 0),$$

which implies directly (3.14). □

Now, for the first part in (3.13) we were able to show that

Proposition 3.13 *There exists $\widehat{C} > 0$ independent of N such that for every $1 \leq f_N < N$*

$$\mathbb{P}\left(\widehat{S}_n(y,N) \geq -f_N,\, n = 1, ..., N-1\right) \leq \widehat{C}\,\frac{(f_N + y)^2}{N}.$$

Proof First we fix $y \geq 0$ and $k < N$ such that $N/k =: L \in \mathbb{N}$. Observe that

$$\left\{S_{kl} - \frac{kl}{N}S_N\right\}_{1 \leq l < L} \stackrel{d}{=} \left\{\sqrt{k}S_l - \frac{kl}{N}\sqrt{k}S_{N/k}\right\}_{1 \leq l < L}.$$

Indeed, since both are centered Gaussian processes it is enough to compare their covariances. Let $l_1 \leq l_2$, $D_l := \sqrt{k}S_l - \frac{kl}{N}\sqrt{k}S_{N/k}$ and $T_l := S_l - \frac{l}{N}S_N$ then

$$\mathrm{Cov}_{\mathbb{P}}(D_{l_1}, D_{l_2}) = \mathbb{E}\left[\left(\sqrt{k}S_{l_1} - \frac{kl_1}{N}\sqrt{k}S_{N/k}\right)\left(\sqrt{k}S_{l_2} - \frac{kl_2}{N}\sqrt{k}S_{N/k}\right)\right]$$

$$= \frac{\sigma^2}{(1-\gamma)^2}\left(kl_1 + \frac{k^2 l_1 l_2}{N^2}k\frac{N}{k} - \frac{kl_1}{N}k\left(l_2 \wedge \frac{N}{k}\right) - \frac{kl_2}{N}k\left(l_1 \wedge \frac{N}{k}\right)\right)$$

$$= \frac{\sigma^2}{(1-\gamma)^2}\left(kl_1 + \frac{k^2 l_1 l_2}{N} - \frac{k^2 l_1 l_2}{N} - \frac{k^2 l_1 l_2}{N}\right)$$

$$= \frac{\sigma^2}{N(1-\gamma)^2}\, kl_1(N - kl_2)$$

3.4 ENTROPIC REPULSION

and

$$\begin{aligned}
\operatorname{Cov}_{\mathbb{P}}(T_{kl_1}, T_{kl_2}) &= \mathbb{E}\left[\left(S_{kl_1} - \frac{kl_1}{N}S_N\right)\left(S_{kl_2} - \frac{kl_2}{N}S_N\right)\right] \\
&= \frac{\sigma^2}{(1-\gamma)^2}\left(kl_1 + \frac{k^2 l_1 l_2}{N^2}N - \frac{kl_1 kl_2}{N} - \frac{kl_2 kl_1}{N}\right) \\
&= \frac{\sigma^2}{N(1-\gamma)^2}\left(kl_1 N + k^2 l_1 l_2 - k^2 l_1 l_2 - k^2 l_1 l_2\right) \\
&= \frac{\sigma^2}{N(1-\gamma)^2} kl_1(N - kl_2) .
\end{aligned}$$

Now setting $n = kl$ and choosing $k := \lfloor f_N + y \rfloor^2$ (of course here k has to be smaller than N, but otherwise our proposition is anyway true for any $\widehat{C} \geq 1$) we can estimate as follows

$$\begin{aligned}
\mathbb{P}\left(\widehat{S}_n(y, N) \geq -f_N,\, n = 1, ..., N-1\right) &\leq \mathbb{P}\left(S_n - \frac{n}{N}S_N \geq -(f_N + y),\, n = 1, ..., N-1\right) \\
&\leq \mathbb{P}\left(S_{kl} - \frac{kl}{N}S_N \geq -(f_N + y),\, 1 \leq kl \leq N-1\right) \\
&= \mathbb{P}\left(\sqrt{k}S_l - \frac{kl}{N}\sqrt{k}S_{N/k} \geq -(f_N + y),\, 1 \leq kl \leq N-1\right) \\
&= \mathbb{P}\left(S_l - \frac{l}{N/k}S_{N/k} \geq -\frac{f_N + y}{\sqrt{k}},\, 1 \leq l \leq \frac{N-1}{k}\right) \\
&\leq \mathbb{P}\left(S_l - \frac{l}{L}S_L \geq -2,\, 1 \leq l \leq L-1\right) \\
&\leq \widehat{C}\,\mathbb{P}\left(S_l - \frac{l}{L}S_L \geq 0,\, 1 \leq l \leq L-1\right) \\
&= \widehat{C}\frac{1}{L} = \widehat{C}\frac{k}{N} \leq \widehat{C}\frac{(f_N + y)^2}{N} .
\end{aligned}$$

Here we have used (3.10) with $y = 0$, Lemma 3.12 and the fact that for a RW started in 0

$$\mathbb{P}\left(S_1 \geq 0, ..., S_{N-1} \geq 0 \mid S_N = 0\right) = \frac{1}{N} ,$$

confer for instance [13]. Of course we have used in the upper estimate that $N/(\lfloor f_N + y \rfloor^2) \in \mathbb{N}$. In the following we will show that this is of no restriction. Let us consider the case where $1 \leq \tilde{L} := \frac{N}{\lfloor f_N + y \rfloor^2} \notin \mathbb{N}$. Here we can find an $1 < c < 2$, such that

$$c\lfloor \tilde{L} \rfloor = \tilde{L} \quad , \text{ or equivalently } \quad c = \frac{\tilde{L}}{\lfloor \tilde{L} \rfloor} .$$

Now, if we set in the upper estimate $k := c\lfloor f_N + y \rfloor^2$, then $L := N/k = \lfloor \tilde{L} \rfloor \in \mathbb{N}$ and we obtain

$$\mathbb{P}\left(\widehat{S}_n(y, N) \geq -f_N,\, n = 1, ..., N-1\right) \leq 2\widehat{C}\frac{(f_N + y)^2}{N} .$$

$\square$

74 The polymer above a "hard wall" : entropic repulsion in Gaussian case

Now by Lemma A.10 there exists an $\hat{c} > 0$ such that

$$r_1(n) - \frac{n}{N} \leq \hat{c} \quad , \text{ for all } N \in \mathbb{N}, n = 1, ..., N-1 \ .$$

Therefore the first part in (3.13) behaves accordingly to Proposition 3.13 like

$$\begin{aligned}
\mathbb{P}\left(\inf_{n=1,...,N-1} \tilde{S}_n(y,N) \geq -f_N\right) &= \mathbb{P}\left(\tilde{S}_n(y,N) \geq -f_N, n = 1, ..., N-1\right) \\
&\leq \mathbb{P}\left(\hat{S}_n(y,N) \geq -(f_N + \hat{c}y), n = 1, ..., N-1\right) \\
&\leq \hat{C} \frac{(f_N + y(1+\hat{c}))^2}{N} \ .
\end{aligned} \quad (3.16)$$

For the second part in (3.13) we have to deal with

$$\hat{U}_n = S_N\left(-\frac{n}{N} + r_1(n) + r_2(n)\right) + U_n - U_N r_1(n) - U_{N+1} r_2(n) \ .$$

Clearly each $\hat{U}_n$ is, as a sum of centered Gaussian variables, also centered Gaussian. Moreover we prove the following

Lemma 3.14 *The variance of $\hat{U}_n$ is uniformly bounded by some constants, i.e.*

$$0 < c_\gamma \leq \text{Var}[\hat{U}_n] \leq C_\gamma \quad , \text{ for all } N \in \mathbb{N}, n = 1, ..., N-1 \ .$$

Proof Let us rewrite $\hat{U}_n$ with the help of 3.12.

$$\begin{aligned}
\hat{U}_n &= \sum_{i=1}^{n} \varepsilon_i \left[-\frac{n}{N} + r_1(n) + r_2(n) + \frac{\gamma^{n-i+1}}{1-\gamma} - \frac{\gamma^{N-i+1}}{1-\gamma} r_1(n) - \frac{\gamma^{N-i+2}}{1-\gamma} r_2(n)\right] \\
&+ \sum_{i=n+1}^{N} \varepsilon_i \left[-\frac{n}{N} + r_1(n) + r_2(n) + 0 - \frac{\gamma^{N-i+1}}{1-\gamma} r_1(n) - \frac{\gamma^{N-i+2}}{1-\gamma} r_2(n)\right] - \varepsilon_{N+1} \frac{\gamma}{1-\gamma} r_2(n) \\
&=: \sum_{i=1}^{n} I_1(i,n,N) + \sum_{i=n+1}^{N} I_2(i,n,N) - \varepsilon_{N+1} \frac{\gamma}{1-\gamma} r_2(n) \ .
\end{aligned}$$

We first attempt the upper bound. By independence, the rough inequality $(a+b+c+d)^2 \leq 16(a^2+b^2+c^2+d^2)$ and the uniform boundedness of $|r_1(n)|, |r_2(n)|$ for

3.4 Entropic repulsion

all $N \in \mathbb{N}, n = 1, ..., N-1$ (cf. Lemma A.10) we obtain

$$\operatorname{Var}[\widehat{U}_n] = \sigma^2 \left(\sum_{i=1}^{n} I_1^2(i,n,N) + \sum_{i=n+1}^{N} I_2^2(i,n,N) + \frac{\gamma^2}{(1-\gamma)^2} r_2^2(n) \right)$$

$$\leq 16\sigma^2 \sum_{i=1}^{n} \left[\left(-\frac{n}{N} + r_1(n) + r_2(n) \right)^2 + \frac{1}{(1-\gamma)^2} \left(\gamma^{2(n-i+1)} + \gamma^{2(N-i+1)} r_1^2(n) + \gamma^{2(N-i+2)} r_2^2(n) \right) \right]$$

$$+ 16\sigma^2 \sum_{i=n+1}^{N} \left[\left(-\frac{n}{N} + r_1(n) + r_2(n) \right)^2 + \frac{1}{(1-\gamma)^2} \left(\gamma^{2(N-i+1)} r_1^2(n) + \gamma^{2(N-i+2)} r_2^2(n) \right) \right]$$

$$+ \sigma^2 \frac{\gamma^2}{(1-\gamma)^2} r_2^2(n)$$

$$\leq \sigma^2 \left(c_1 + 16 \sum_{i=1}^{N} \left(-\frac{n}{N} + r_1(n) + r_2(n) \right)^2 \right), \tag{3.17}$$

where we have used finiteness of the geometric sum above and c_1 is some positive constant. We will show now that the last expression is bounded by a constant. First of all one can compute (cf. Appendix A.5)

$$\sup_{n=1,...,N-1} N(r_1(n) + r_2(n))$$

$$= \sup_{n=1,...,N-1} N \frac{-n + \gamma(1-\gamma^n + n) + \gamma^{N-n+1}(1+\gamma^n(-1+(-1+(-1+\gamma)n)))}{-N + \gamma(2+N+\gamma^N(-2+(-1+\gamma)N))}$$

$$\underset{N \to \infty}{\sim} \sup_{n=1,...,N-1} \frac{-n + \gamma(1-\gamma^n + n) + \gamma^{N-n+1}}{-1+\gamma} = \sup_{n=1,...,N-1} \left(n + \frac{\gamma(1-\gamma^n) + \gamma^{N-n+1}}{-1+\gamma} \right). \tag{3.18}$$

Therefore we conclude

$$\sup_{n=1,...,N-1} \sum_{i=1}^{N} \left(-\frac{n}{N} + r_1(n) + r_2(n) \right)^2 = \sup_{n=1,...,N-1} \frac{1}{N} (-n + (r_1(n) + r_2(n))N)^2$$

$$\sim \sup_{n=1,...,N-1} \frac{1}{N} \left[\frac{\gamma(1-\gamma^n) + \gamma^{N-n+1}}{-1+\gamma} \right]^2 \xrightarrow[N \to \infty]{} 0.$$

Now from 3.17 we conclude that there is a constant C_γ, such that the upper bound in the Lemma is satisfied. Next we prove the lower bound, but this is easy, since from the first equation in 3.17 it follows that

$$\operatorname{Var}[\widehat{U}_n] \geq \frac{\sigma^2}{(1-\gamma)^2} \sum_{i=1}^{n} \gamma^{2(n-i+1)} \geq \sigma^2 \frac{\gamma^2}{(1-\gamma)^2} =: c_\gamma > 0.$$

$\square$

Now setting $f_N = \tilde{c}\sqrt{\log N}$ for an $\tilde{c} > 0$ we can conclude with Lemma 3.14

$$\mathbb{P}\left(\sup_{n=1,\ldots,N-1} |\widehat{U}_n| \geq f_N\right) \leq \sum_{i=1}^{N-1} \mathbb{P}\left(|\widehat{U}_n| \geq f_N\right) \leq 2N \sup_{i=1,\ldots,N-1} \mathbb{P}\left(\widehat{U}_n \geq f_N\right)$$

$$\leq \frac{2N}{\sqrt{2\pi c_\gamma}} \frac{e^{-\frac{f_N^2}{2C_\gamma}}}{f_N}$$

$$\leq \sqrt{\frac{2}{\pi c_\gamma}} \frac{1}{\tilde{c}\sqrt{\log N}} N^{-\left(\frac{\tilde{c}^2}{2C_\gamma}-1\right)}. \tag{3.19}$$

We choose $\tilde{c} > 0$ such that

$$\frac{\tilde{c}^2}{2C_\gamma} - 1 \geq 1 \quad, \text{ i.e. } \quad \tilde{c} \geq 2\sqrt{C_\gamma}. \tag{3.20}$$

Finally we arrive by a quite sharp upper bound given here in the

Proposition 3.15 *There exist constants $c, C > 0$ and $N_0 \in \mathbb{N}$, s. th. for all $x, y \geq 0$ and $N \geq N_0$*

$$w_{x,y}(N) \leq C \frac{\log N}{N}(1+cy)^2,$$

Proof The proof is now just a collection of our previous results. By (3.7) and Lemma 3.10 we have for fixed $x, y \geq 0$ and $N \geq N_0$ (for some $N_0 \in \mathbb{N}$):

$$w_{x,y}(N) = \mathrm{P}^{(-x,0)}\left(\widehat{\Omega}^+_{N-2}\right) \leq \mathbb{P}\left(\widehat{\Omega}^+_{N-2}\right).$$

Furthermore by the decoupling (3.13) and the estimates (3.16) and (3.19) there exist some constants $\hat{c}, \widehat{C}, \bar{c} > 0$ and $p \geq 1$, such that

$$w_{x,y}(N) \leq \widehat{C} \frac{(f_{N-1} + y(1+\hat{c}))^2}{N-1} + \frac{\bar{c}}{\sqrt{N-1}(N-1)^p}.$$

Therefore recalling $f_N = \tilde{c}\sqrt{\log N}$ with $\tilde{c}$ chosen like in (3.20) and setting $C := 4\max\{\widehat{C}\tilde{c}^2, \bar{c}\}$ we estimate further on with $c := 1 + \hat{c}$

$$w_{x,y}(N) \leq \widehat{C} \frac{f_{N-1}^2(1+cy)^2}{N-1} + \frac{\bar{c}}{\sqrt{N-1}(N-1)^p}$$

$$= \widehat{C}\tilde{c}^2 \frac{\log N-1}{N-1}(1+cy)^2 + \frac{\bar{c}}{\sqrt{N-1}(N-1)^p}$$

$$\leq \frac{C}{4}\left(\frac{\log N-1}{N-1} + \frac{1}{\sqrt{N-1}(N-1)^p}\right)(1+cy)^2$$

$$\leq \frac{C}{2} \frac{\log N-1}{N-1}(1+cy)^2 \leq C\frac{\log N}{N}(1+cy)^2.$$

□

3.4.2 Lower bound

In proving the lower bound for the conditional orthant probability in Definition 3.5, we restrict ourselves to the case $x = y = 0$, which will be sufficient for our purposes. First we show an usefull inclusion bound. Recall therefore the definition of the Gaussian i.i.d. sequence $\{\varepsilon_i\}_{i \in \mathbb{Z}^+}$ in subsection 3.3.1.

Lemma 3.16 *Let $\tilde{S}_0 := 0$, $\tilde{S}_n := \varepsilon_1 + ... + \varepsilon_n$ and $\Lambda_N^+ := \{\tilde{S}_1 \geq 0, ..., \tilde{S}_N \geq 0\}$. Then for all $N \in \mathbb{N}$ and $\varepsilon > 0$ we have*

$$\{\Lambda_N^+, 0 \leq Y_N \leq \varepsilon\} \supseteq \{\Lambda_N^+, \tilde{S}_N \leq \varepsilon\},$$

where $Y_0 = 0$.

Proof Recall that $Y_n = \gamma^{n-1}\varepsilon_1 + ... + \gamma^0 \varepsilon_n$ with an $0 < \gamma < 1$. We prove the Lemma by induction. Since $\tilde{S}_1 = Y_1 = \varepsilon_1$ the statement for $n = 1$ is trivial. Now for an fixed $n \in \mathbb{N}$ let for all $\varepsilon > 0$

$$\{\Lambda_N^+, 0 \leq Y_N \leq \varepsilon\} \supseteq \{\Lambda_N^+, \tilde{S}_N \leq \varepsilon\}. \tag{3.21}$$

It is then

$$\Lambda_{N+1}^+ \cap \{0 \leq \tilde{S}_{N+1} \leq \varepsilon\} = \{\Lambda_N^+\} \cap \{\tilde{S}_N \leq \varepsilon - \varepsilon_{N+1}\} \cap \{0 \leq \tilde{S}_{N+1}\}$$
$$\stackrel{(3.21)}{\subseteq} \{\Lambda_N^+\} \cap \{0 \leq Y_N \leq \varepsilon - \varepsilon_{N+1}\} \cap \{0 \leq \tilde{S}_{N+1}\}$$
$$\subseteq \{\Lambda_{N+1}^+\} \cap \{0 \leq \gamma Y_N \leq \varepsilon - \varepsilon_{N+1}\}$$
$$= \{\Lambda_{N+1}^+\} \cap \{0 \leq Y_{N+1} \leq \varepsilon\}.$$

□

Remark 3.17 *Observe that of course $S_n = \tilde{S}_n/(1-\gamma)$ (cf. (3.9)) and therefore the last Lemma also applies to "our" random walk $\{S_n\}$ instead of $\{\tilde{S}_n\}$.*

The result of this subsection is the following lower bound.

Proposition 3.18 $w_{0,0}(N)$ *has a polynomial lower bound in N.*

Proof Proposition 1.8 allows us to write

$$w_{0,0}(N) = \mathrm{P}^{(0,0)}(\Omega_{N-2}^+ |\ W_{N-1} = 0,\ W_N = 0) = \frac{1}{\varphi_N^{(0,0)}(0,0)} \mathcal{Z}_{0,N-1}^w \frac{1}{\lambda^N}.$$

For the sake of convenience we want to consider just an odd number for N. The reason is, that now we have an even number of field variables and it is possible to use a symmetry argument. In this case the boundary conditions are $\varphi_{-1} = \varphi_0 = \varphi_{2N} = \varphi_{2N+1} = 0$ and by Remark 1.13

$$w_{0,0}(2N+1) \geq c_1 \frac{\sqrt{2N+1}}{\lambda^{2N+1}} \int_{\mathbb{R}_+^{2N-1}} e^{-\sum_{i=1}^{2N+1} V_1(\nabla \varphi_i) - \sum_{i=0}^{2N} V_2(\Delta \varphi_i)} \prod_{i=1}^{2N-1} d\varphi_i\ ,$$

where $V_1(\eta) = \eta^2\alpha/2$ and $V_2(\eta) = \eta^2\beta/2$. Since we want to obtain a lower bound, we restrict the integration to $C_N^1(\varepsilon) := \mathbb{R}_+^{2N-1} \cap \{|\varphi_N - \varphi_{N-1}| < \varepsilon, |\varphi_N - \varphi_{N+1}| < \varepsilon\}$, for an arbitrary fixed $\varepsilon > 0$. On $C_N^1(\varepsilon)$ we have $|\nabla\varphi_{N+1}| < \varepsilon$ and $|\Delta\varphi_N| < 2\varepsilon$. Trivially there is an $M_\varepsilon > 0$ with $x^2\beta/2 \leq M_\varepsilon$ for all $|x| < 2\varepsilon$. Therefore we obtain

$$w_{0,0}(2N+1)/(c_1\sqrt{2N+1})$$
$$\geq \frac{e^{-M_\varepsilon}}{\lambda^{2N+1}} \int_{C_N^1(\varepsilon)} e^{-\sum_{i=1}^N V_1(\nabla\varphi_i) - \sum_{i=0}^{N-1} V_2(\Delta\varphi_i)} \, e^{-\sum_{i=N+1}^{2N} V_1(\nabla\varphi_i) - \sum_{i=N+1}^{2N} V_2(\Delta\varphi_i)} \prod_{i=1}^{2N-1} d\varphi_i$$
$$= \frac{e^{-M_\varepsilon}}{\lambda^{2N+1}} \int_{\mathbb{R}_+} d\varphi_N \left[\int_{C_N^2(\varepsilon)} e^{-\sum_{i=1}^N V_1(\nabla\varphi_i) - \sum_{i=0}^{N-1} V_2(\Delta\varphi_i)} \prod_{i=1}^{N-1} d\varphi_i \right]^2,$$

where in the last step we have used the symmetry of the quadratic potential $x^2\alpha/2$ and the symmetry of the integrand on $C_N^1(\varepsilon)$, setting $C_N^2(\varepsilon) := \mathbb{R}_+^{N-1} \cap \{|\varphi_N - \varphi_{N-1}| < \varepsilon\}$. Now we take $c_N > 0$ and make a further restriction on the integration, then we use Jensen's inequality

$$w_{0,0}(2N+1)/(c_1\sqrt{2N+1})$$
$$\geq \frac{e^{-M_\varepsilon}}{\lambda^{2N+1}} \int_0^{c_N} d\varphi_N \left[\int_{C_N^2(\varepsilon)} e^{-\sum_{i=1}^N V_1(\nabla\varphi_i) - \sum_{i=0}^{N-1} V_2(\Delta\varphi_i)} \prod_{i=1}^{N-1} d\varphi_i \right]^2$$
$$\geq \frac{e^{-M_\varepsilon}}{c_N \lambda} \left[\frac{1}{\lambda^N} \int_0^{c_N} d\varphi_N \int_{C_N^2(\varepsilon)} e^{-\sum_{i=1}^N V_1(\nabla\varphi_i) - \sum_{i=0}^{N-1} V_2(\Delta\varphi_i)} \prod_{i=1}^{N-1} d\varphi_i \right]^2$$
$$= \frac{e^{-M_\varepsilon}}{c_N \lambda} \left[\frac{1}{\lambda^N} \int_0^{c_N} d\varphi_N \int_{C_N^2(\varepsilon)} \frac{\nu(0)}{\nu(\varphi_N - \varphi_{N-1})} \right.$$
$$\left. \cdot \frac{\nu(\varphi_N - \varphi_{N-1})}{\nu(0)} e^{-\sum_{i=1}^N V_1(\nabla\varphi_i) - \sum_{i=0}^{N-1} V_2(\Delta\varphi_i)} \prod_{i=1}^{N-1} d\varphi_i \right]^2$$
$$\geq \frac{e^{-M_\varepsilon}}{c_N \lambda} \left[P^{(0,0)}(\Omega_N^+, W_N \leq c_N, |W_N - W_{N-1}| \leq \varepsilon) \right]^2, \quad (3.22)$$

because $\nu(x) \leq 1$ for $\beta \geq 0$ and in the last inequality we have used Proposition 1.8. Now by Lemma 3.16 and the remark after

$$P^{(0,0)}(\Omega_N^+, W_N \leq c_N, |W_N - W_{N-1}| \leq \varepsilon) \geq P^{(0,0)}(\Lambda_N^+, W_N \leq c_N, |W_N - W_{N-1}| \leq \varepsilon)$$
$$\geq P^{(0,0)}(\Lambda_N^+, |Y_N| \leq \varepsilon) - P^{(0,0)}(W_N > c_N)$$
$$\geq P^{(0,0)}(\Lambda_N^+, 0 \leq Y_N \leq \varepsilon) - P^{(0,0)}(W_N > c_N)$$
$$\geq P^{(0,0)}(\Lambda_N^+, S_N \leq \varepsilon) - P^{(0,0)}(W_N > c_N). \quad (3.23)$$

Using a duality Lemma and a combinatorial identity of Alili and Doney [2], in [10] it was shown that for the first term one can obtain

$$P^{(0,0)}(\Lambda_N^+, S_N \leq \varepsilon) \geq \frac{1}{N} P^{(0,0)}(S_N \in [0, \varepsilon]) \sim \frac{\text{const.}}{N^{3/2}}.$$

3.5 IMPACT OF PINNING IN WETTING-MODEL

Now, the goal is to obtain a sufficiently small upper bound on $\mathrm{P}^{(0,0)}(W_N > c_N)$ to estimate (3.22) polynomially from below. For this reason we set $c_N = N^c$ with an $c > 0$, which we will have to specify later, and consider by Markov-inequality

$$\mathrm{P}^{(0,0)}(W_N > N^c) \leq \frac{1}{N^c} E_{\mathrm{P}^{(0,0)}} |W_N| \leq \frac{1}{N^c} \sum_{i=1}^{N} E_{\mathrm{P}^{(0,0)}} |Y_i| \leq \frac{1}{N^c} \sum_{i=1}^{N} \sum_{j=1}^{i} E_{\mathrm{P}^{(0,0)}} |\gamma^{i-j} \varepsilon_j|$$

$$\leq \frac{1}{N^c} E_{\mathrm{P}^{(0,0)}} |\varepsilon_1| \sum_{i=1}^{N} \sum_{j=1}^{i} \gamma^{i-j} \leq \frac{1}{N^c} E_{\mathrm{P}^{(0,0)}} |\varepsilon_1| \frac{N}{1-\gamma}$$

$$= \frac{E_{\mathrm{P}^{(0,0)}} |\varepsilon_1|}{1-\gamma} \frac{1}{N^{c-1}} < \infty .$$

Since we can choose $c_N > 0$ arbitrarily, we can take $c > 0$ as large as we want and therefore by (3.23) there exists a constant such that

$$\mathrm{P}^{(0,0)}(\Omega_N^+, W_N \leq c_N , |W_N - W_{N-1}| \leq \varepsilon) \geq \frac{\mathrm{const.}}{N^{3/2}} .$$

In particular this works for $c = 2$ and so by (3.22) we can finally state

$$w_{0,0}(2N+1) \geq c_0 \frac{e^{-M_\varepsilon}}{\lambda} \frac{(\mathrm{const.})^2}{N^5} \sqrt{2N+1} \geq c_0 \frac{e^{-M_\varepsilon}}{\lambda} \frac{(\mathrm{const.})^2}{N^{9/2}} .$$

□

3.5 Impact of pinning in wetting-model

So far only the free wetting model $\mathbb{P}_{0,N}^w$ was studied, i.e. only the self-interaction and the repelling property, coming from the presence of a wall, have been taken into account. Now it is time to approach a description when an additional "strength" attracts the chain at the defect-line. This describes exactly the model $\mathbb{P}_{\varepsilon,N}^w$ for an $\varepsilon > 0$. In what follows we will have several similar steps to the pinning case. In order to make it readable we will translate them into the setting of wetting case, but try to avoid unnecessary repetitions.

3.5.1 The contact process

We define the contact process $(\tau_i)_{i \in \mathbb{Z}^+}$ by

$$\tau_0 := 0 \quad \text{and} \quad \tau_{i+1} := \inf\{k > \tau_i \,|\, \varphi_k = 0\}$$

and the process $(J_i)_{i \in \mathbb{Z}^+}$, which gives the height of the polymer before the contact points

$$J_0 := 0 \quad \text{and} \quad J_i := \varphi_{\tau_i - 1} .$$

Set the contact number as usual $\ell_N = \#\{i \in \{1, ..., N\} \,|\, \varphi_i = 0\}$ and take for fixed $k \in \mathbb{N}$ a time-partition $(t_i)_{i=1,...,k} \in \mathbb{N}$ with $0 < t_1 < \cdots < t_{k-1} < t_k := N$. For $(y_i)_{i=1,...,k} \in \mathbb{R}$

the joint law of the process $\{\ell_N, (\tau_i)_{i \leq \ell_N}, (J_i)_{i \leq \ell_N}\}$ is

$$\mathbb{P}^w_{\varepsilon,N}(\ell_N = k, \tau_i = t_i, J_i \in dy_i, \; i = 1, ..., k)$$
$$= \frac{\varepsilon^{k-1}}{Z^w_{\varepsilon,N}} F^w_{0,dy_1}(t_1) F^w_{y_1,dy_2}(t_2 - t_1) \cdots F^w_{y_{k-1},dy_k}(N - t_{k-1}) F^w_{y_k,\{0\}}(1) \,, \qquad (3.24)$$

where $F^w_{x,dy}(n) := f^w_{x,y}(n)\mu(dy)$, $\mu(dy) := \delta_0(dy) + dy$ and

$$f^w_{x,y}(n) := \begin{cases} e^{-\beta x^2/2} \mathbb{1}_{\{y=0\}} \mathbb{1}_{\{x,y \geq 0\}} & , n = 1 \\ e^{-\mathcal{H}_{[-1,2]}(x,0,y,0)} \mathbb{1}_{\{y \neq 0\}} \mathbb{1}_{\{x,y \geq 0\}} & , n = 2 \\ \int_{\mathbb{R}^{n-2}_+} e^{-\mathcal{H}_{[-1,n]}(w_{-1},...,w_n)} \mathbb{1}_{\{y \neq 0\}} \mathbb{1}_{\{x,y \geq 0\}} dw_1 \cdots dw_{n-2} & , n \geq 3 \\ \text{with } w_{-1} = x, w_0 = 0, w_{n-1} = y, w_n = 0 \,. \end{cases} \qquad (3.25)$$

Next we set

$$\widetilde{f}^w_{x,y}(n) := \frac{\nu(-y)}{\lambda^n \nu(-x)} f^w_{x,y}(n) \quad \text{and} \quad \widetilde{F}^w_{x,dy}(n) := \widetilde{f}^w_{x,y}(n) \mu(dy) \,.$$

For $n \geq 3$ and $x, y \in \mathbb{R}$ with $w_{-1} = x, w_0 = 0, w_{n-1} = y, w_n = 0$ we can write with $f_{.,.}(.)$ from (1.24)

$$f^w_{x,y}(n) = \int_{\mathbb{R}^{n-2}} e^{-\mathcal{H}_{[-1,n]}(w)} \prod_{i=1}^{n-2} dw_i \; \frac{\int_{\mathbb{R}^{n-2}_+} e^{-\mathcal{H}_{[-1,n]}(w)} \prod_{i=1}^{n-2} dw_i}{\int_{\mathbb{R}^{n-2}} e^{-\mathcal{H}_{[-1,n]}(w)} \prod_{i=1}^{n-2} dw_i} \mathbb{1}_{\{y \neq 0\}} \mathbb{1}_{\{x,y \geq 0\}} = f_{x,y}(n) \, w_{x,y}(n) \qquad (3.26)$$

and therefore with the help of the density (1.8)

$$\widetilde{f}^w_{x,y}(n) := \frac{\nu(-y)}{\lambda^n \nu(-x)} f_{x,y}(n) \, w_{x,y}(n) = \varphi^{(-x,0)}_n(y,0) \, w_{x,y}(n) \,.$$

3.5.2 Construction of a semi Markov (sub)-kernel

We are going to describe the process of contact points in a more exact way. Our model consists of three-body interaction terms, therefore it won't be possible to describe $(\tau_i)_{i \in \mathbb{Z}^+}$ by a renewal process. Nevertheless something else can be proven, but first we define

$$K^{w,\varepsilon}_{x,dy}(n) := \varepsilon \widetilde{F}^w_{x,dy}(n) e^{-F^w_s(\varepsilon)n} \frac{\nu^w_\varepsilon(y)}{\nu^w_\varepsilon(x)} \,, \qquad (3.27)$$

for an F^w_s and ν^w, which are preliminary specified in the next proposition.

Proposition 3.19 *There exists an $\varepsilon^w_c \in (0, \infty)$ such that for every $\varepsilon \in (0, \infty)$ there exist $F^w_s(\varepsilon) \in [0, \infty)$ and $\nu^w(\varepsilon) \in (0, \infty)$ with the property*

$$\int_{y \in \mathbb{R}} \sum_{n \in \mathbb{N}} K^{w,\varepsilon}_{x,dy}(n) = \min\left\{\frac{\varepsilon}{\varepsilon^w_c}, 1\right\} \,, \text{ for all } x \in \mathbb{R} \,. \qquad (3.28)$$

We postpone the proof to subsection 3.6.2, where a lot more can be said about F_s^w and ν^w and even an explicit representation can be given.

This proposition is very useful, because it says that $K_{\cdot,\cdot}^{w,\varepsilon}(\cdot)$ denotes just a semi Markov (sub-)kernel. The probabilistic interpretation of such kernels is the fact that one can now define a law $\mathcal{P}_\varepsilon^w$ under which $\{(\tau_i, J_i)\}_{i \in \mathbb{Z}^+}$ is a (defective for $0 < \varepsilon < \varepsilon_c^w$) Markov chain on $\mathbb{Z}^+ \times \mathbb{R}$ with $(\tau_0, J_0) = (0,0)$ and the transition kernel

$$\mathcal{P}_\varepsilon^w((\tau_{i+1}, J_{i+1}) \in (\{n\}, dy) \,|\, (\tau_i, J_i) = (m,x)) = K_{x,dy}^{w,\varepsilon}(n-m) \ . \tag{3.29}$$

Then the contact process $(\tau_i)_{i \in \mathbb{Z}^+}$ is called a Markov renewal process and $(J_i)_{i \in \mathbb{Z}^+}$ its modulating chain. Furthermore we can rewrite (3.24) as follows

$$\mathbb{P}_{\varepsilon,N}^w(\ell_N = k, \tau_i = t_i, J_i \in dy_i,\ i=1,...,k)$$
$$= \frac{e^{F_s^w(\varepsilon)(N+1)}}{\varepsilon^2 \mathcal{Z}_{\varepsilon,N}^w} \lambda^{N+1} K_{0,dy_1}^{w,\varepsilon}(t_1) K_{y_1,dy_2}^{w,\varepsilon}(t_2 - t_1) \cdots K_{y_{k-1},dy_k}^{w,\varepsilon}(N - t_{k-1}) K_{y_k,\{0\}}^{w,\varepsilon}(1) \tag{3.30}$$

and for $t_0 = y_0 = 0$ the normalizing constant of $\mathbb{P}_{\varepsilon,N}^w$ has then to be

$$\mathcal{Z}_{\varepsilon,N}^w = \frac{e^{F_s^w(\varepsilon)(N+1)}}{\varepsilon^2} \lambda^{N+1} \sum_{k=1}^N \sum_{\substack{t_i \in \mathbb{N}, i=1,...,k \\ 0 < t_1 < \cdots < t_k := N}} \int_{\mathbb{R}^k} \left(\prod_{i=1}^k K_{y_{i-1},dy_i}^{w,\varepsilon}(t_i - t_{i-1}) \right) K_{y_k,\{0\}}^{w,\varepsilon}(1) \ . \tag{3.31}$$

The next result reveals a connection between $\mathbb{P}_{\varepsilon,N}^w$, which is dependent on N, and $\mathcal{P}_\varepsilon^w$, which is not.

Proposition 3.20 *Define $\mathcal{A}_N := \{\exists j \geq 0 \,|\, \tau_j = N, \tau_{j+1} = N+1\}$. Then for all $N \in \mathbb{N}, \varepsilon > 0$ and $k \leq N$ ($(t_i)_{i=1,...,k}, (y_i)_{i=1,...,k}$ as usual)*

$$\mathbb{P}_{\varepsilon,N}^w(\ell_N = k, \tau_i = t_i, J_i \in dy_i,\ i \leq k) = \mathcal{P}_\varepsilon^w(\ell_N = k, \tau_i = t_i, J_i \in dy_i,\ i \leq k \,|\, \mathcal{A}_N)$$

and

$$\mathcal{Z}_{\varepsilon,N}^w = \frac{e^{F_s^w(\varepsilon)(N+1)}}{\varepsilon^2} \lambda^{N+1} \mathcal{P}_\varepsilon^w(\mathcal{A}_N) \ . \tag{3.32}$$

Proof Due to (3.29) we have

$$\mathcal{P}_\varepsilon^w(\ell_N = k, \tau_i = t_i, J_i \in dy_i,\ i < k \,|\, \mathcal{A}_N)$$
$$= \frac{1}{\mathcal{P}_\varepsilon^w(\mathcal{A}_N)} K_{0,dy_1}^{w,\varepsilon}(t_1) K_{y_1,dy_2}^{w,\varepsilon}(t_2 - t_1) \cdots K_{y_{k-1},dy_k}^{w,\varepsilon}(N - t_{k-1}) K_{y_k,\{0\}}^{w,\varepsilon}(1)$$

and knowing that $\mathcal{P}_\varepsilon^w(\cdot \,|\, \mathcal{A}_N)$ is a probability measure and comparing with (3.30) and (3.31) we arrive at the end of the proof. □

3.6 Accurate determination of F_s^w

In this section we are going to proof Proposition 3.19 and give explicit representations for the quantities F_s^w and ν^w.

3.6.1 An useful bound and the Hilbert-Schmidt property

Let us introduce for every $\theta \geq 0$ and $h \in L^2(\mathbb{R}, d\mu)$ the operator

$$(B^{w,\theta} h)(x) := \int_{\mathbb{R}} B^{w,\theta}_{x,dy} h(y) \quad , \text{ where } B^{w,\theta}_{x,dy} := \sum_{n \in \mathbb{N}} e^{-\theta n} \widetilde{F}^w_{x,dy}(n) . \qquad (3.33)$$

We are going to see later that this operator is a Hilbert-Schmidt operator on the Hilbert-space $L^2(\mathbb{R}, d\mu)$. As a first step we prove the following Lemma, which uses the the explicit representation (3.5) of the integrated Markov chain.

Lemma 3.21 *For all $x, y \geq 0$ and $n \in \mathbb{N}_{\geq 2}$ there exists $c_1, c_2, c_3 > 0$ such that*

$$\widetilde{f}_{x,y}(n) \leq \frac{c_1}{n^{1/2}} \exp\left\{-\frac{1}{2}\left(c_2 y^2 + \frac{c_3}{n} x^2\right)\right\} .$$

Proof Recall the representation (1.21) for $n \geq 2$

$$\widetilde{f}_{x,y}(n) = \varphi_n^{(-x,0)}(y, 0) \mathbb{1}_{\{y \neq 0\}}$$
$$= \frac{1}{2\pi \sqrt{\det(\Sigma_n)}} \exp\left\{-\frac{1}{2}\left\langle \begin{pmatrix} y - \mu_{n-1}^{\alpha,\beta}(-x,0) \\ -\mu_n^{\alpha,\beta}(-x,0) \end{pmatrix}, \Sigma_n^{-1} \begin{pmatrix} y - \mu_{n-1}^{\alpha,\beta}(-x,0) \\ -\mu_n^{\alpha,\beta}(-x,0) \end{pmatrix}\right\rangle\right\} \mathbb{1}_{\{y \neq 0\}} ,$$

where

$$\Sigma_n = \begin{pmatrix} \text{Cov}(W_{n-1}, W_{n-1}) & \text{Cov}(W_{n-1}, W_n) \\ \text{Cov}(W_n, W_{n-1}) & \text{Cov}(W_n, W_n) \end{pmatrix}$$

and

$$\mu_n^{\alpha,\beta}(-x, 0) = E_{\mathrm{P}^{(-x,0)}} W_n = -x\gamma \frac{1-\gamma^n}{1-\gamma} .$$

First of all we know already from (1.15) that

$$\frac{\sqrt{n}}{2\pi \sqrt{\det(\Sigma_n)}} \longrightarrow \text{const.} > 0 \quad , \ n \to \infty .$$

Now the canonical scalar-product above can be written as

$$y^2 a(\gamma, n) + \frac{x^2 \gamma^2}{(1-\gamma)^2} b(\gamma, n) + \frac{2xy\gamma}{1-\gamma} c(\gamma, n) . \qquad (3.34)$$

We will prove successively that a, b and c can be estimated in such a way that the Lemma is fulfilled. We set

$$\Sigma_n^{-1} =: \begin{pmatrix} H_{1,1}(n) & H_{1,2}(n) \\ H_{2,1}(n) & H_{2,2}(n) \end{pmatrix}$$

3.6 Accurate determination of F_s^w

where one can calculate from the representation (3.5) (cf. Appendix A.5)

$$H_{1,1}(n) = \frac{(-1+\gamma)(-n+\gamma(2+\gamma+\gamma^{1+2n}-2\gamma^n(1+\gamma)+\gamma n))}{(-1+\gamma^n)(1+\gamma+\gamma^n(-1+\gamma(-1+n))-n)-n+\gamma n)\sigma^2}$$

$$H_{1,2}(n) = H_{2,1}(n) = \frac{(-1+\gamma)(-1-\gamma^{1+2n}+\gamma^n(1+\gamma)^2+n-\gamma(1+\gamma+\gamma n))}{(-1+\gamma^n)(1+\gamma+\gamma^n(-1+\gamma(-1+n))-n)-n+\gamma n)\sigma^2}$$

$$H_{2,2}(n) = \frac{(-1+\gamma)(1+\gamma^{2n}-2\gamma^n(1+\gamma)-n+\gamma(2+\gamma n))}{(-1+\gamma^n)(1+\gamma+\gamma^n(-1+\gamma(-1+n))-n)-n+\gamma n)\sigma^2} \;.$$

It is easily seen that

$$H_{1,1}(n) \xrightarrow[n\to\infty]{} \frac{(-1+\gamma)(-1+\gamma^2)}{-(-1+\gamma)\sigma^2} = \frac{1-\gamma^2}{\sigma^2} > 0$$

$$H_{1,2}(n) = H_{2,1}(n) \xrightarrow[n\to\infty]{} \frac{(-1+\gamma)(1-\gamma^2)}{-(-1+\gamma)\sigma^2} = -\frac{1-\gamma^2}{\sigma^2} < 0 \quad (3.35)$$

$$H_{2,2}(n) \xrightarrow[n\to\infty]{} \frac{(-1+\gamma)(-1+\gamma^2)}{-(-1+\gamma)\sigma^2} = \frac{1-\gamma^2}{\sigma^2} > 0 \;.$$

Let us take a look at a in (3.34). Since Σ_n is positive definite, $a(\gamma, n) = H_{1,1}(n) > 0$ for all $n \in \mathbb{N}_{\geq 2}$. Taking (3.35) into account, a can be bounded from below by a constant, say $c_2 > 0$, which will only depend on γ of course.

Next by positive definiteness, definition and calculation we obtain

$$0 < b(\gamma, n) = \left\langle \begin{pmatrix} 1-\gamma^{n-1} \\ 1-\gamma^n \end{pmatrix}, \Sigma_n^{-1} \begin{pmatrix} 1-\gamma^{n-1} \\ 1-\gamma^n \end{pmatrix} \right\rangle$$

$$= \frac{(-1+\gamma)^3(-\gamma^2+2\gamma^{1+n}(1+\gamma)+\gamma^{2n}(-1-n+\gamma^2(-1+(-2+\gamma)\gamma+n)))}{\gamma^2(-1+\gamma^{1+n})(-n+\gamma(2+n+\gamma^n(-2+(-1+\gamma)n)))\sigma^2}$$

and therefore

$$n\,b(\gamma, n) \xrightarrow[n\to\infty]{} \frac{(-1+\gamma)^3(-\gamma^2)}{-\gamma^2(-1+\gamma)\sigma^2} = \frac{(1-\gamma)^2}{\sigma^2} > 0 \;.$$

This means we can find an $c_3 > 0$ such that

$$\frac{x^2\gamma^2}{(1-\gamma)^2} b(\gamma, n) \geq \frac{c_3}{n} \;.$$

Finally we will show that $c(\gamma, n) \geq 0$ for all $n \in \mathbb{N}$. For this observe that by definition and calculation

$$c(\gamma, n) = H_{1,1}(n)(1-\gamma^{n-1}) + H_{1,2}(n)(1-\gamma^n)$$

$$= \frac{(-1+\gamma)^2(\gamma-\gamma^{1+2n}+\gamma^n(-1+\gamma^2)n)}{\gamma(-1+\gamma^n)(1+\gamma+\gamma^n(-1+\gamma(-1+n))-n)-n+\gamma n)\sigma^2}$$

$$=: \frac{(-1+\gamma)^2(\gamma-\gamma^{1+2n}+\gamma^n(-1+\gamma^2)n)}{\gamma(-1+\gamma^n)d_1(\gamma,n)\sigma^2} \quad (3.36)$$

$$\sim \frac{1-\gamma}{n\sigma^2} > 0 \;.$$

We will show that the positivity even holds for every $n \in \mathbb{N}$. Now $d_1(\gamma, 1) = 0$ and we have the equivalence

$$d_1(\gamma, n+1) < d_1(\gamma, n) \tag{3.37}$$
$$\Longleftrightarrow 1 + \gamma + \gamma^{n+1}(-1 + \gamma n - (n+1)) - (n+1) + \gamma(n+1)$$
$$< 1 + \gamma + \gamma^n(-1 + \gamma(-1+n) - n) - n + \gamma n$$
$$\Longleftrightarrow \gamma^n - 1 + \gamma - \gamma^{n+1} - n(2\gamma^{n+1} - \gamma^{n+2} - \gamma^n) < 0$$
$$\Longleftrightarrow (-1 + \gamma)(1 - \gamma^n + n\gamma^n(\gamma - 1)) < 0$$
$$\Longleftrightarrow : (-1 + \gamma) d_2(\gamma, n) < 0$$

The term d_2 has to be non-negative for all $0 \leq \gamma \leq 1$ and $n \in \mathbb{N}$ because

$$\frac{\partial}{\partial \gamma} d_2(\gamma, n) = -\gamma^{n-1} n + \gamma^n n + (\gamma - 1)\gamma^{n-1} n^2$$

and so

$$\frac{\partial}{\partial \gamma} d_2(\gamma, n) = 0 \Leftrightarrow n\gamma^{n-1}(\gamma - 1)(n+1) = 0 \Leftrightarrow \gamma = 0 \text{ or } \gamma = 1 \ .$$

These are the only possible extrema and so by observing that $d_2(0, n) = 1$, $d_2(1, n) = 0$ one can see the non-negativeness of d_2. Moreover this implies the monotony in (3.37) and so $d_1(g, n+1) < d_1(\gamma, 1) = 0$ for all $n \in \mathbb{N}$. Therfore by (3.36) we have the following equivalence

$$c(\gamma, n) > 0$$
$$\Longleftrightarrow \gamma - \gamma^{1+2n} + \gamma^n(-1 + \gamma^2)n > 0 \ ,$$

but

$$\gamma - \gamma^{1+2n} + \gamma^n(-1 + \gamma^2)n \geq -\gamma^{1+2n} + \gamma^{n+2}n = \gamma^{n+2}(n - \gamma^{n-1}) > 0$$

and therefore indeed $c(\gamma, n) > 0$ for all $n \in \mathbb{N}$. To obtain the stated result in the Lemma we can now estimate (3.34) from below by taking simply zero instead of c. □

Now we are ready to tackle the Hilbert-Schmidt property.

Proposition 3.22 *For every $\theta \geq 0$ the operator $B^{w,\theta}$ defined in (3.33) is a Hilbert-Schmidt operator on the Hilbert-space $L^2(\mathbb{R}, d\mu)$.*

Proof Let $\theta \geq 0$. We set $B^{w,\theta}_{x, dy} = b^{w,\theta}(x, y)\mu(dy)$ and

$$b^{w,\theta}(x, y) := e^{-\theta} \widetilde{f}^w_{x,0}(1) \mathbb{1}_{\{y=0\}} + \sum_{n \geq 2} e^{-\theta n} \widetilde{f}^w_{x,y}(n) \mathbb{1}_{\{y \neq 0\}} \ ,$$

then we have to show

$$\int_{\mathbb{R}} \int_{\mathbb{R}} b^{w,\theta}(x, y)^2 \, \mu(dx) \, \mu(dy) \ < \infty \ .$$

It is

$$b^{w,\theta}(x, y)^2 = e^{-2\theta} \widetilde{f}^w_{x,0}(1)^2 \mathbb{1}_{\{y=0\}} + \sum_{n, m \geq 2} e^{-\theta(n+m)} \widetilde{f}^w_{x,y}(n) \widetilde{f}^w_{x,y}(m) \mathbb{1}_{\{y \neq 0\}}$$

3.6 Accurate determination of F_s^w

and setting $\theta = 0$ we can estimate from above

$$\int_{\mathbb{R}} \int_{\mathbb{R}} b^{w,\theta}(x,y)^2 \, \mu(dx) \, \mu(dy) \leq \int_{\mathbb{R}} \tilde{f}_{x,0}^w(1)^2 \, \mu(dx) + \sum_{n,m \geq 2} \int_{\mathbb{R}} \tilde{f}_{0,y}^w(n) \tilde{f}_{0,y}^w(m) \, dy$$
$$+ \sum_{n,m \geq 2} \int_{\mathbb{R}} \int_{\mathbb{R}} \tilde{f}_{x,y}^w(n) \tilde{f}_{x,y}^w(m) \, dx \, dy \; .$$

The first term on the r.h.s.

$$\tilde{f}_{x,0}^w(1)^2 = e^{-\beta x^2} \left(\frac{\nu(0)}{\lambda \nu(-x)} \right)^2 \mathbb{1}_{\{x \geq 0\}} = \left(\frac{\nu(0)}{\lambda} \right)^2 \exp\left(-\frac{x^2}{2} \left[2\beta + \alpha - \sqrt{\alpha}\sqrt{\alpha + 4\beta} \right] \right) \mathbb{1}_{\{x \geq 0\}}$$

is integrable for our conditions (**AP**) on α and β, because [...] > 0 (cf. Calculation A.9). Let us take $n \geq 2$. From Lemma 3.21 we know that $\tilde{f}_{0,y}(n) \leq c_1/\sqrt{n}$ and by Proposition 3.15 we obtain for $n, m \geq N_0$

$$\int_{\mathbb{R}} \tilde{f}_{0,y}^w(n) \tilde{f}_{0,y}^w(m) \, dy \leq C c_1 \frac{\log n}{n^{3/2}} \int_{\mathbb{R}} (1+cy)^2 \tilde{f}_{0,y}^w(m) \, dy$$
$$\leq C^2 c_1^2 \frac{\log(n) \log(m)}{(nm)^{3/2}} \int_{\mathbb{R}_+} (1+cy)^4 e^{-\frac{c_2}{2} y^2} \, dy < \infty$$

and this is obviously summable. Similarly one sees that there is also no problem with summation over $\mathbb{N}_{\geq 2}^2 \setminus \{(n,m) \, | \, n, m \geq N_0\}$. So the second term is also all right.
For the last term we use a similar calculation using first a symmetry argument

$$\sum_{n,m \geq N_0} \int_{\mathbb{R}} \int_{\mathbb{R}} \tilde{f}_{x,y}^w(n) \tilde{f}_{x,y}^w(m) \, dx \, dy \leq 2 \sum_{n \geq N_0} \sum_{m \geq n} \int_{\mathbb{R}} \int_{\mathbb{R}} \tilde{f}_{x,y}^w(n) \tilde{f}_{x,y}^w(m) \, dx \, dy$$
$$\leq 2C^2 c_1^2 \sum_{n \geq N_0} \sum_{m \geq n} \frac{\log(n) \log(m)}{(nm)^{3/2}} \int_{\mathbb{R}_+} \int_{\mathbb{R}_+} (1+cy)^4 e^{-\frac{1}{2}(c_2 y^2 + \frac{c_3}{n} x^2)} \, dx \, dy$$
$$\leq k_1 \sum_{n \geq N_0} \sum_{m \geq n} \frac{\log(n) \log(m)}{(nm)^{3/2}} n^{1/2}$$

for some $k_1 > 0$. Now there exists an constant $k_2 > 0$ such that $\log m \leq k_2 m^{1/4}$ for all $m \geq N_0$ and therefore

$$\sum_{m \geq n} \frac{\log(m)}{m^{3/2}} \leq k_2 \sum_{m \geq n} m^{-5/4} \leq \tilde{k}_2 n^{-1/4},$$

for some $\tilde{k}_2 > 0$. Taking this into account we obtain finally

$$\sum_{n,m \geq 2} \int_{\mathbb{R}} \int_{\mathbb{R}} \tilde{f}_{x,y}^w(n) \tilde{f}_{x,y}^w(m) \, dx \, dy \leq k_1 \tilde{k}_2 \sum_{n \geq N_0} \frac{\log n}{n^{5/4}} + C < \infty \; ,$$

where again due to symmetry

$$C := 2 \sum_{n=2}^{N_0 - 1} \sum_{m \geq n} \int_{\mathbb{R}} \int_{\mathbb{R}} \tilde{f}_{x,y}^w(n) \tilde{f}_{x,y}^w(m) \, dx \, dy < \infty$$

and so we have finished the proof. $\square$

3.6.2 Perron-Frobenius Thm. and proof of Proposition 3.19

By Proposition 3.22 we have shown in particular the compactness of $B^{w,\theta}$ on $L^2(\mathbb{R}, d\mu)$. Thus we can apply an infinite dimensional Perron-Frobenius theorem of Zerner, cf Appendix A.1. For this purpose let $\theta \geq 0$ and $\delta^w(\theta) \in (0, \delta(0)]$ be the spectral radius of the operator $B^{w,\theta}$. Just like for the pinning case it can be shown that $\delta^w(.)$ is strictly decreasing, but now on $[0, \infty)$. So we consider the inverse $(\delta^w)^{-1}$ on $(0, \delta^w(0)]$ and

$$\varepsilon_c^w := \frac{1}{\delta^w(0)}, \quad F_s^w(\varepsilon) := \begin{cases} (\delta^w)^{-1}(1/\varepsilon) & , \text{for } \varepsilon \geq \varepsilon_c^w \\ 0 & , \text{for } \varepsilon \leq \varepsilon_c^w \end{cases} \quad (3.38)$$

The symbol F_s^w is of course already reserved for the notion of the free energy, but we will see later that both indeed coincide. For $\varepsilon \geq 0$ we consider the operator $B^{w,F_s^w(\varepsilon)}$ with its spectral radius

$$\delta^w(F_s^w(\varepsilon)) = \min\left\{\frac{1}{\varepsilon}, \frac{1}{\varepsilon_c^w}\right\}$$

The kernel of $B^{w,F_s^w(\varepsilon)}$ is strictly positive, so Zerner's theorem ensures the existence of the right and left Perron-Frobenius eigenfunctions $\nu_\varepsilon^w(.), w_\varepsilon^w(.) \in L^2(\mathbb{R}, d\mu)$, such that $\nu_\varepsilon^w(x), w_\varepsilon^w(x) > 0$ for μ-a.e. $x \in \mathbb{R}$ and $\nu_\varepsilon^w(x) = w_\varepsilon^w(x) = 0$ for all $x < 0$. This means

$$\int_{y \in \mathbb{R}} B_{x,dy}^{w,F_s^w(\varepsilon)} \nu_\varepsilon^w(y) = \min\left\{\frac{1}{\varepsilon}, \frac{1}{\varepsilon_c^w}\right\} \nu_\varepsilon^w(x) \quad (3.39)$$

and

$$\int_{x \in \mathbb{R}} w_\varepsilon^w(x) B_{x,dy}^{w,F_s^w(\varepsilon)} \mu(dx) = \min\left\{\frac{1}{\varepsilon}, \frac{1}{\varepsilon_c^w}\right\} w_\varepsilon^w(y) \mu(dy). \quad (3.40)$$

From this one even sees that $\nu_\varepsilon^w(x), w_\varepsilon^w(x) > 0$ for all $x \geq 0$.

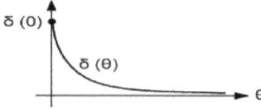

Figure 3.2: A sketch of the spectral radius $\delta(.)$. It is strcitly decreasing with $0 < \delta(0) < \infty$ and $\lim_{\theta \to \infty} \delta(\theta) = 0$.

Proof of Proposition 3.19

We were not very precise in using the same notation for ε_c^w, F_s^w and ν_ε^w like the one in Proposition 3.19, since it is yet not clear if they satisfy what we would like to have. Nevertheless the lines above tell us that the only remaining thing about those candidates is to prove (3.28), but using (3.39) this is indeed true

$$\int_{y\in\mathbb{R}} \sum_{n\in\mathbb{N}} K^{w,\varepsilon}_{x,dy}(n) = \frac{\varepsilon}{\nu^w_\varepsilon(x)} \int_{y\in\mathbb{R}} \left(\sum_{n\in\mathbb{N}} \tilde{F}^w_{x,dy}(n) e^{-F^w_s(\varepsilon)n} \right) \nu^w_\varepsilon(y)$$

$$= \frac{\varepsilon}{\nu^w_\varepsilon(x)} \int_{y\in\mathbb{R}} B^{w,F^w_s(\varepsilon)}_{x,dy} \nu^w_\varepsilon(y)$$

$$= \varepsilon \min\left\{\frac{1}{\varepsilon}, \frac{1}{\varepsilon^w_c}\right\} = \min\left\{1, \frac{\varepsilon}{\varepsilon^w_c}\right\}.$$

□

Remark 3.23 *According to (3.29) and Proposition 3.19, the process $(J_i)_{i\in\mathbb{Z}^+}$ is a Markov chain on $\mathbb{R}$. It is defective for $0 \leq \varepsilon < \varepsilon^w_c$. The chain starts in $J_0 = 0$ and has the transition kernel*

$$\mathcal{P}^w_\varepsilon(J_{i+1} \in dy \,|\, J_i = x) = \sum_{n\in\mathbb{N}} K^{w,\varepsilon}_{x,dy}(n) =: D^{w,\varepsilon}_{x,dy}.$$

The left and right eigenfunctions are defined up to multiplicative constant, so we can assume from now on that $\langle \nu^w_\varepsilon, w^w_\varepsilon \rangle = \int_\mathbb{R} \nu^w_\varepsilon w^w_\varepsilon \, d\mu = 1$. This means $\kappa^w_\varepsilon(dx) := \nu^w_\varepsilon(x) w^w_\varepsilon(x) \mu(dx)$ is a probability measure on $\mathcal{B}(\mathbb{R})$. Due to (3.39), if $\varepsilon \geq \varepsilon^w_c$ then κ^w_ε is invariant for $D^{w,\varepsilon}_{x,dy}$:

$$\int_{x\in\mathbb{R}} D^{w,\varepsilon}_{x,dy} \kappa^w_\varepsilon(dx) = \int_{x\in\mathbb{R}} \left(\sum_{n\in\mathbb{N}} \tilde{F}^w_{x,dy}(n) e^{-F^w_s(\varepsilon)n} \right) \frac{\varepsilon}{\nu^w_\varepsilon(x)} \nu^w_\varepsilon(y) \nu^w_\varepsilon(x) w^w_\varepsilon(x) \mu(dx)$$

$$= \varepsilon \nu^w_\varepsilon(y) \int_{x\in\mathbb{R}} B^{w,F^w_s(\varepsilon)}_{x,dy} w^w_\varepsilon(x) \mu(dx) = \kappa^w_\varepsilon(dy).$$

Therefore $(J_i)_{i\in\mathbb{Z}^+}$ is a positive recurrent Markov chain under $\mathcal{P}^w_\varepsilon$, if $\varepsilon \geq \varepsilon^w_c$, cf. [25].

3.7 Identification of the free energy and proof of Thm. 3.1

In this section we will prove the localization-delocalization result, which was stated in Theorem 3.1. In particular we will see the connection of previous results to the free energy.

3.7.1 Obtaining an ordinary renewal process

We have already seen, that $(\tau_i)_{i\in\mathbb{Z}^+}$ is a Markov renewal process. In what follows we need a "sub-process" of $(\tau_i)_{i\in\mathbb{Z}^+}$, which will be a classical (i.e. non Markov) renewal process. Namely, we define the double-contact process $(\eta_i)_{i\in\mathbb{Z}^+}$ by

$$\eta_0 := 0 \quad, \quad \eta_{i+1} := \inf\{k > \eta_i \,|\, \varphi_{k-1} = \varphi_k = 0\}.$$

Because of the special structure of the transition kernel (3.29) and the remark 3.23 the following proposition of [10] applies.

Proposition 3.24 *The process $(\eta_i)_{i\in\mathbb{Z}^+}$ under $\mathcal{P}^w_\varepsilon$ is a classical renewal process, which is non terminating for $\varepsilon \geq \varepsilon^w_c$.*

3.7.2 Proof of Theorem 3.1

We will first show that the wetting model displays a (non-trivial) phase transition. To obtain this, it remains to show is that the expression F_s^w defined in (3.38) for all $\varepsilon \geq 0$ indeed coincides with the free energy

$$F^w(\varepsilon) = \lim_{N \to \infty} \frac{1}{N} \log \left(\frac{\mathcal{Z}_{\varepsilon,N}^w}{\mathcal{Z}_{0,N}^w} \right) . \tag{3.41}$$

Before we attempt this, let us consider a small, but important

Lemma 3.25 *The free partition function from the wetting case has the following limiting behavior*

$$\frac{1}{n} \log \mathcal{Z}_{0,N}^w \xrightarrow[N \to \infty]{} \log \lambda ,$$

where λ is the spectral radius of the compact operator K defined in 1.3

Proof It is by definition 3.5 and similarly to (3.26)

$$\mathcal{Z}_{0,N}^w = \mathcal{Z}_{0,N} \, w_{0,0}(N+1)$$

and in (1.10) we have already proven

$$\frac{1}{N} \log \mathcal{Z}_{0,N} \xrightarrow[N \to \infty]{} \log \lambda .$$

Thus by the upper- and lower bound on $w_{\cdot,\cdot}(\cdot)$, i.e. Proposition 3.15 and Proposition 3.18, we obtain also

$$\lim_{N \to \infty} \frac{1}{N} \log \mathcal{Z}_{0,N}^w = \log \lambda + \lim_{N \to \infty} \frac{1}{N} \log w_{0,0}(N+1) = \log \lambda .$$

$\square$

Now, with the help of (3.32) we can write for $\varepsilon > 0$

$$\frac{\mathcal{Z}_{\varepsilon,N}^w}{\mathcal{Z}_{0,N}^w} = \frac{e^{F_s^w(\varepsilon)(N+1)}}{\varepsilon^2 \mathcal{Z}_{0,N}^w} \lambda^{N+1} \mathcal{P}_\varepsilon^w(\mathcal{A}_N)$$

and so

$$\frac{1}{N} \log \left(\frac{\mathcal{Z}_{\varepsilon,N}^w}{\mathcal{Z}_{0,N}^w} \right) = \frac{N+1}{N} F_s^w(\varepsilon) + \frac{N+1}{N} \log \lambda + \frac{1}{N} \log \mathcal{P}_\varepsilon^w(\mathcal{A}_N) - \frac{2}{N} \log \varepsilon - \frac{1}{N} \log \mathcal{Z}_{0,N}^w . \tag{3.42}$$

Due to Lemma 3.25, in the limit $N \to \infty$ we can neglect the second and last term on the right hand side. In view of Definition 3.38 there are two cases to distinguish between, namely $0 \leq \varepsilon \leq \varepsilon_c^w$ and $\varepsilon > \varepsilon_c^w$.

3.7 IDENTIFICATION OF THE FREE ENERGY AND PROOF OF THM. 3.1

Let us consider the first one $0 \leq \varepsilon \leq \varepsilon_c^w$. From the monotonicity of the partition function in ε and the fact that $\mathcal{P}_\varepsilon^w$ is a probability measure we can easily estimate

$$0 \leq \lim_{N\to\infty} \frac{1}{N} \log\left(\frac{\mathcal{Z}_{\varepsilon,N}^w}{\mathcal{Z}_{0,N}^w}\right) \leq F_s^w(\varepsilon) + \lim_{N\to\infty} \frac{1}{N} \log \mathcal{P}_\varepsilon^w(\mathcal{A}_N) \leq F_s^w(\varepsilon) = 0 \ .$$

The last equality is just by definition (3.38) for $0 \leq \varepsilon \leq \varepsilon_c^w$.
We turn to the second case $\varepsilon > \varepsilon_c^w$, in which, again by definition (3.38), $F_s^w(\varepsilon) > 0$. Considering (3.42), to complete the identification of the free energy it remains to check that one has

$$\lim_{N\to\infty} \frac{1}{N} \log \mathcal{P}_\varepsilon^w(\mathcal{A}_N) = 0 \ . \quad (3.43)$$

The set $\mathcal{A}_N$, defined in proposition 3.20, can be written as $\mathcal{A}_N = \{\exists j \geq 0 \,|\, \eta_j = N+1\}$. It is known that (3.43) is true for any non-terminating aperiodic renewal process, cf. [18] Theorem A.3. However $(\eta_i)_{i\in\mathbb{Z}^+}$ is aperiodic, because

$$\mathcal{P}_\varepsilon^w(\eta_1 = 1) = \mathcal{P}_\varepsilon^w((\tau_1, J_1) \in (\{1\}, \{0\}) \,|\, (\tau_0, J_0) = (0,0)) = K_{0,\{0\}}^{w,\varepsilon}(1) = \frac{\varepsilon}{\lambda} e^{-F_s^w(\varepsilon)} > 0$$

and due to proposition 3.24 it is a non-terminating renewal process under $\mathcal{P}_\varepsilon^w$ for $\varepsilon > \varepsilon_c^w$. Altogether we have shown

$$F_s^w(\varepsilon) = F^w(\varepsilon) = \lim_{N\to\infty} \frac{1}{N} \log\left(\frac{\mathcal{Z}_{\varepsilon,N}^w}{\mathcal{Z}_{0,N}^w}\right) \quad , \varepsilon \geq 0 \ .$$

We have already studied the property of analyticity in the localized regime $\mathcal{L}$ before we defined F_s^w. Finally what is left is the asymptotic behavior of the free energy $F^w(\varepsilon)$ as $\varepsilon \to \infty$. The idea is to use a sandwich argument and therefore first of all by (3.33) consider

$$e^{-\theta} \tilde{F}_{x,dy}^w(1) \leq B_{x,dy}^{w,\theta} \leq e^{-\theta} B_{x,dy}^{w,0} \ . \quad (3.44)$$

Further on one can consider their corresponding integral operators on $L^2(\mathbb{R}, d\mu)$, e.g. $(\tilde{\mathcal{F}}h)(x) := \int \tilde{F}_{x,dy}^w(1)h(y)$. In particular it is by (3.25) for $h(x) = e^{-V^{(2)}(x)}/\nu(-x)$:

$$(\tilde{\mathcal{F}}h)(x) = \int \tilde{F}_{x,y}^w(1) h(y) \, \delta_0(dy) = \frac{1}{\lambda} h(x) \ .$$

Now the same inequality as in (3.44) has to be valid for the spectral radius of $B^{w,\theta}$ and $B^{w,0}$, this means

$$\frac{1}{\lambda} e^{-\theta} \leq \delta^w(\theta) \leq e^{-\theta} \delta^w(0) \ .$$

Now recalling $\delta^w(0) \in (0, \infty)$ and setting $\theta := (\delta^w)^{-1}(1/\varepsilon)$ we obtain

$$\log\left(\frac{\varepsilon}{\lambda}\right) \leq F^w(\varepsilon) \leq \log(\varepsilon \, \delta^w(0)) \ ,$$

which implies the asymptotic behavior in Theorem 3.1.

4 Phase Transitions for higher dimensional Gaussian models

We would like to extend the results from the $(1+1)$-dimensional case to the $(1+d)$-dimensional one. Since in higher dimensions also a pinning subspace of $\mathbb{R}^d$ can be chosen in different ways, it is sensible to ask whether another behavior than up to now occurs.

4.1 The model in higher dimensions

The following model is a generalization of our $(1+1)$-dimensional model for a linear chain with pinning and wetting effect. On the one hand this generalization is motivated by a paper from Bolthausen, Funaki and Otobe [7], where they introduce a pinning measure ν and certain pinning subspaces. On the other hand $\mathcal{H}_N$ in (4.2) is a modification for our purposes of the Hamiltonian appearing in a paper by Sakagawa [27] or [22] for $(d+1)$-dimensional Gaussian free models. Although this Hamiltonian is written in a quite general way, we will concentrate on case (iii), see next page, and make some comments on other cases. Now let us consider the distribution of the chain, given by the polymer measure on $(\mathbb{R}^d)^{N-1}$:

$$\mathbb{P}^{(+)}_{\varepsilon,N,d,m}(d\varphi_1,...,d\varphi_{N-1}) := \frac{\exp(-\mathcal{H}_N(\varphi))}{\mathcal{Z}^{(+)}_{\varepsilon,N,d,m}} \prod_{i=1}^{N-1} \left(\varepsilon\nu(d\varphi_i^{(+)}) + d\varphi_i^{(+)}\right), \qquad (4.1)$$

where

- $\varepsilon \geq 0$ is the usual pinning parameter
- $+$ denotes the model with an additional wall, i.e. $\varphi_n \in \mathbb{R}^{d-1} \times \mathbb{R}_+$ for all n
- $d\varphi_i^{(+)}$ the Lebesgue measure on $\mathbb{R}^d$ or $\mathbb{R}^{d-1} \times \mathbb{R}_+$ respectively
- $\mathcal{Z}^{(+)}_{\varepsilon,N,d,m}$ the normalization constant (partition function)
- The Hamiltonian $\left(\text{for some fixed } K \in \mathbb{N} \text{ and } u_j := \lfloor\frac{j+1}{2}\rfloor + \lfloor\frac{j}{2}\rfloor - 1\right)$

$$\mathcal{H}_N(\varphi) = \mathcal{H}_{[-K+1,N+K-1]}(\varphi) = \sum_{j=1}^{K} q_j \sum_{i=-\lfloor\frac{j}{2}\rfloor+1}^{N+\lfloor\frac{j+1}{2}\rfloor-1} V\left((-\Delta)^{j/2}\varphi_i\right) \qquad (4.2)$$

and we take zero boundary conditions

$$\varphi_{-u_K} = ... = \varphi_0 = \varphi_N = ... = \varphi_{N+u_K} = 0$$

- The potential $V : \mathbb{R}^d \to \mathbb{R}$, $\eta \mapsto \|\eta\|_2^2/2$
- Let M be an m-dimensional subspace of $\mathbb{R}^d$ for $0 \leq m \leq d-1$ (for the model with the wall we take $M \subset \partial(\mathbb{R}^{d-1} \times \mathbb{R}_+))$.
- The pinning measure on $\mathbb{R}^d$ is $\nu(dy) := dy^{(1)}\delta_0(dy^{(2)})$, where $dy^{(1)}$ is the Lebesgue measure on M and the second measure is the Dirac mass at 0 on $M^\perp$.

By the definition it is $(-\Delta)^{j/2}\varphi_n = (-\Delta)^{\lfloor j/2 \rfloor}\nabla\varphi_n$ and so by induction we can easily see

$$(-\Delta)^{j/2}\varphi_n = \sum_{k=0}^{j}\binom{j}{k}(-1)^k \varphi_{i+\lfloor j/2 \rfloor - k}$$

and therefore the Hamiltonian (4.2) has the form

$$H_N(\varphi) = \sum_{j=1}^{K} q_j \sum_{i=-\lfloor\frac{i-2}{2}\rfloor}^{N+\lfloor\frac{i-1}{2}\rfloor} V\left(\sum_{k=0}^{j}\binom{j}{k}(-1)^k \varphi_{i+\lfloor j/2 \rfloor - k}\right).$$

Although the Hamiltonian in our model is written in a quite general way, we will focus on the following three cases:

(i) $q_1 \neq 0, q_2 = ... = q_K = 0 \rightsquigarrow$ the gradient case
(ii) $q_2 \neq 0, q_1 = q_3 = ... = q_K = 0 \rightsquigarrow$ the Laplacian case
(iii) $q_1, q_2 \neq 0, q_3 = ... = q_K = 0 \rightsquigarrow$ (our) mixed case.

The gradient case is well known and was studied for instance in [7] and [17] for higher dimensions and Gaussian potentials. The Laplacian case was studied by Caravenna and Deuschel [10] in $(1+1)$-dimensions, but for more general potentials. Finally, the mixed case we have studied already in the previous chapters in $(1+1)$-dimensions.
Now we will extend our model to higher dimensions and Gaussian potentials. We will also make some statements on the higher dimensional Laplacian model.

From now on let us consider the mixed case, i.e. case (iii) with $q_1 = \alpha/2$ and $q_2 = \beta/2$.

4.2 Free partition function

The first observation is one due to the quadratic potential V. We can write the free partition function ($\varepsilon = 0$) in the following way

$$\mathcal{Z}_{0,N,d,m}^{(+)} = (\mathcal{Z}_{0,N,1,0})^{d-1} \mathcal{Z}_{0,N,1,0}^{(+)}. \tag{4.3}$$

Of course the free partition function is not dependent on the pinning subspace and so we omit here the notation of the last parameter m. We remark that from the equation above and earlier results

$$\lim_{N \to \infty} \frac{1}{N} \log \mathcal{Z}_{0,N,d}^{(+)} = (d-1) \lim_{N \to \infty} \frac{1}{N} \log \mathcal{Z}_{0,N,1} + \lim_{N \to \infty} \frac{1}{N} \log \mathcal{Z}_{0,N,1}^{(+)} = d \log \lambda \ ,$$

with λ from Proposition 1.5.

Remark 4.1
With (4.3) and the results known so far we obtain the following asymptotical behavior of $\mathcal{Z}_{0,N,d}^{(+)}/\lambda^{Nd}$

	pinning	wetting		
∇-case	$O(N^{-d/2})$	$O(N^{-(d+2)/2})$		
Δ-case	$O(N^{-2d})$	$O(N^{-2d}) = \mathcal{Z}_{0,N,d} \geq \mathcal{Z}_{0,N,d}^{+} \geq \frac{const.}{N^{2+c}} \mathcal{Z}_{0,N,1}^{d-1} = \frac{const.}{N^{2d+c}}$	, $c > 0$
mixed-case	$O(N^{-d/2})$	$\mathcal{Z}_{0,N,d}^{+} \leq c \cdot \log N / N^{(d+2)/2}$ (conj.: same as ∇-case)		

In a "pure" case like the gradient or Laplacian, one usually takes $\lambda = 1$, i.e. a "normalized" potential. For the Gaussian case it would be $V(\eta) = \pi |\eta|^2$ and so one can for instance easily see that the free energy at the origin is zero. In contrast, the free energy of mixed models has to be normalized, cf. previous chapters, in order to have $F(0) = 0$.

Remark 4.2
In the table above we can see that the mixed model has the same asymptotical behavior as the gradient model. We conjecture that this fact is in general true, meaning that asymptotical behavior of a model is always determined by the term in (4.2) with the lowest non-vanishing $k = \inf\{i \geq 1 | q_i \neq 0\}$. Therefore it behaves like the corresponding "pure"-model (only q_k is non-zero), for which we conjecture from several computations for the "normalized" potential $V(\eta) = \pi |\eta|^2$

$$\mathcal{Z}_{0,N,d} \sim \frac{c_k}{N^{k^2 d/2}} \ , \quad \text{where } k = \inf\{i \geq 1 | q_i \neq 0\}$$

and c_k is just a constant depending on k.

4.3 First example: heavy pinning

A simple case is what we call the heavy pinning, i.e. $M = \{x \in \mathbb{R}^d | x_d = 0\}$. By the name heavy pinning we want to indicate that the pinning occurs on the greatest subspace we want to consider. In this case $\nu(dy) := dy^{(1)} \delta_0(dy^{(2)})$, where $dy^{(1)}$ is the Lebesgue measure on M and the second measure is the Dirac mass at 0 on $M^\perp = \mathbb{R}$ or $M^\perp = \mathbb{R}_+$ for wetting.

For the integration over $(\mathbb{R}^d)^{N-1}$ we can write the product of measures in (4.1) as

$$\prod_{i=1}^{N-1}(\varepsilon\nu(d\varphi_i^{(+)})+d\varphi_i^{(+)}) = \prod_{i=1}^{N-1}\left(\varepsilon d\varphi_i^{(1,\ldots,d-1)}\delta_0(d\varphi_i^{(d)})+d\varphi_i\right)$$
$$= \prod_{i=1}^{N-1} d\varphi_i^{(1,\ldots,d-1)}\left(\varepsilon\delta_0(d\varphi_i^{(d)})+d\varphi_i^{(d)}\right)$$
$$= \left(\prod_{i=1}^{N-1} d\varphi_i^{(1)}\cdots d\varphi_i^{(d-1)}\right)\left(\prod_{i=1}^{N-1}(\varepsilon\delta_0(d\varphi_i^{(d)})+d\varphi_i^{(d)})\right) \quad (4.4)$$

where $(1,\ldots,d-1)$ in the superscript denotes the first $d-1$ components of φ_i, i.e. $\varphi_i^{(1,\ldots,d-1)} \in \mathbb{R}^{d-1}$. Therefore the partition function can be written as the product

$$\mathcal{Z}_{\varepsilon,N,d,d-1}^{(+)} = (\mathcal{Z}_{0,N,1,0})^{d-1}\mathcal{Z}_{\varepsilon,N,1,0}^{(+)}. \quad (4.5)$$

Recalling equation (4.3) we can state

$$F_{d,d-1}^{(+)}(\varepsilon) = \lim_{N\to\infty}\frac{1}{N}\log\left(\frac{\mathcal{Z}_{\varepsilon,N,d,d-1}^{(+)}}{\mathcal{Z}_{0,N,d,d-1}^{(+)}}\right) = \lim_{N\to\infty}\frac{1}{N}\log\left(\frac{\mathcal{Z}_{\varepsilon,N,1,0}^{(+)}}{\mathcal{Z}_{0,N,1,0}^{(+)}}\right) = F^{(+)}(\varepsilon),$$

where $F^{(+)}(\varepsilon)$ is the free energy of the corresponding $(1+1)$-dimensional model. So we obtain the

Proposition 4.3 (Heavy pinning)
The heavy pinning models in $(1+d)$-dimensions possess exactly the same free energy formula like their corresponding $(1+1)$-dimensional analogies.

This means in particular that the localization behavior, analyticity, order of phase transition, etc. remain the same. This is of course what one expects, because compared to the state space, we have here a very strong pinning subspace $M = \mathbb{R}^{d-1}$. Crucial, as we will see later, is the difference between the dimension of the state space of φ_i and $\dim(M)$ and here it is one.

Remark 4.4
The Proposition above is of course not only valid for our mixed model, but for all Gaussian models having the product-measure in (4.1).

4.4 The weak pinning

Having seen the heavy pinning, we go from the one extreme to the other one, which we call weak pinning, i.e. $M = \{0\}$. In this case $\nu(dy) = \delta_0(dy)$ is the Dirac mass at 0 on $\mathbb{R}^d$. Here the representation of the partition function like in (4.5) is only possible for $\varepsilon = 0$.

4.4 THE WEAK PINNING

We choose the same approach like in chapter 1 and 3 to obtain results for localization-delocalization in higher dimensions. In order to avoid all the recapitulation and reformulation into the higher dimensional case, we will concentrate on the substantial differences here. We think that having read these previous chapters one is enabled to understand what will go on. The proof for the mixed model in $(1+1)$-dimensions can be more or less adapted and goes through also for $d \geq 2$. Nevertheless, the step where we have to be careful is the Hilbert-Schmidt property in Lemma 1.16 or 3.22. Crucial for the proof is the compactness of B^θ.

Let us recall briefly some definitions. We will state them only in the pinning case, because it is similar for the wetting case and one has to restrict to the state space $\mathbb{R}^{d-1} \times \mathbb{R}_+$. For $x \in \mathbb{R}^d$ we define

$$(B^\theta h)(x) := \int_\mathbb{R} B^\theta_{x,dy} h(y) , \qquad (4.6)$$

where

$$B^\theta_{x,dy} := \sum_{n \in \mathbb{N}} e^{-\theta n} \widetilde{F}_{x,dy}(n)$$

and for $x, y \in \mathbb{R}^d$

$$\widetilde{F}_{x,dy}(n) := \widetilde{f}_{x,y}(n) \mu(dy) \quad \text{and} \quad \widetilde{f}_{x,y}(n) := \frac{\nu(-y)}{\lambda^{dn} \nu(-x)} f_{x,y}(n) .$$

The kernels $f_{.,.}(.)$ have the form

$$f_{x,y}(n) := \begin{cases} e^{-\beta \|x\|_2^2/2} \mathbb{1}_{\{y=0\}} & , n = 1 \\ e^{-\mathcal{H}_{[-1,2]}(x,0,y,0)} \mathbb{1}_{\{y \neq 0\}} & , n = 2 \\ \int_{\mathbb{R}^{n-2}} e^{-\mathcal{H}_{[-1,n]}(w_{-1},\ldots,w_n)} dw_1 \cdots dw_{n-2} \mathbb{1}_{\{y \neq 0\}} & , n \geq 3 \\ \text{with } w_{-1} = x, w_0 = 0, w_{n-1} = y, w_n = 0 . \end{cases}$$

We state the following

Proposition 4.5
For all $x, y \in \mathbb{R}^d$ and $n \in \mathbb{N}$

(a) $f_{x,y}(n) = \prod_{i=1}^d \left(f_{x^{(i)},y^{(i)}}(n) \right)$, where $x = (x^{(1)}, \cdots, x^{(d)})^T$

(b) $\nu(x) := \prod_{i=1}^d \left(\nu(x^i) \right)$ is the strict positive right-eigenfunction coresponding to the spectral radius λ^d of the operator K.

Here again we recall the definition of K, for $f \in L^2(\mathbb{R}^d, dx)$

$$K(x, f) := \int f(y) k(x, y) dy$$

and $k(x, y) = e^{-\frac{\alpha}{2} \|y\|_2^2 - \frac{\beta}{2} \|y-x\|_2^2} .$

Proof Of course the usage of the same notation for $f_{.,.}(.)$ and $\nu(.)$ is somewhat misleading, but we don't want to use more indices. The statements above are trivial, for instance we see for each $x \in \mathbb{R}^d$

$$K(x,\nu) = \int_{\mathbb{R}^d} k(x,y)\nu(y)\,dy = \int_{\mathbb{R}^d} \prod_{i=1}^d e^{-\frac{\alpha}{2}|y^i|^2 - \frac{\beta}{2}|y^{(i)} - x^{(i)}|^2} \nu(y^{(i)})\,dy$$

$$= \prod_{i=1}^d \int_{\mathbb{R}} e^{-\frac{\alpha}{2}|y^i|^2 - \frac{\beta}{2}|y^{(i)} - x^{(i)}|^2} \nu(y^{(i)})\,dy^{(i)} = \prod_{i=1}^d \lambda \nu(x^{(i)}) = \lambda^d \nu(x)$$

The strict positivity is due to that from the $(1+1)$-dimensional case. □

In the pinning case, we will show the compactness of the operator B^θ for $\theta \geq 0$ on

- $L^2(\mathbb{R}^d, d\mu)$, if $d \geq 5$ and
- $L^1(\mathbb{R}^d, d\mu)$, if $d = 3$ or $d = 4$.

The case $d = 1$ was already investigated before. Like in $d = 1$, we will see that in dimension $d = 2$ the operator B^θ is only compact iff $\theta > 0$.
In the wetting case it will be shown that $B^{\theta,+}$, $\theta \geq 0$ is compact on $L^2(\mathbb{R}^d, d\mu)$ for all $d \geq 1$.

4.4.1 Compactness for wetting and high-dimensional pinning

We first turn to the higher dimensional pinning case, i.e. we consider dimensions greater or equal than five. Furthermore also wetting in all dimensions will be considered.

Proposition 4.6
In the pinning case for every $d \geq 5$ and $\theta \geq 0$ the operator B^θ is compact on the Hilbert-space $L^2(\mathbb{R}^d, d\mu)$. Whereas the same holds for $B^{\theta,+}$ in the wetting case, but for all $d \geq 1$.

Proof In what follows we refer to Lemma 1.16 or 3.22. Here we show for the kernel of $B^{\theta,(+)}$

$$b^{\theta,(+)}(x,y) := e^{-\theta} \widetilde{f}_{x,0}^{(+)}(1) \mathbb{1}_{\{y=0\}} + \sum_{n \geq 2} e^{-\theta n} \widetilde{f}_{x,y}^{(+)}(n) \mathbb{1}_{\{y \neq 0\}}$$

the stronger condition

$$\int_{\mathbb{R}^d} \int_{\mathbb{R}^d} b^\theta(x,y)^2 \,\mu(dx)\,\mu(dy) < \infty \,, \tag{4.7}$$

i.e. the Hilbert-Schmidt property. We know already that for all $\theta \geq 0$

$$\int_{\mathbb{R}^d} \int_{\mathbb{R}^d} b^{\theta,(+)}(x,y)^2 \,\mu(dx)\,\mu(dy) \leq \int_{\mathbb{R}^d} \widetilde{f}_{x,0}^{(+)}(1)^2 \,\mu(dx) + \sum_{n,m \geq 2} \int_{\mathbb{R}^d} \widetilde{f}_{0,y}^{(+)}(n) \widetilde{f}_{0,y}^{(+)}(m)\,dy$$

$$+ \sum_{n,m \geq 2} \int_{\mathbb{R}^d} \int_{\mathbb{R}^d} \widetilde{f}_{x,y}^{(+)}(n) \widetilde{f}_{x,y}^{(+)}(m)\,dx\,dy \,, \tag{4.8}$$

where

$$\widetilde{f}_{x,y}^{(+)}(n) = \frac{\nu(-y)}{\lambda^{dn}\nu(-x)} f_{x,y}^{(+)}(n) = \frac{\nu(-y)}{\lambda^{dn}\nu(-x)} \prod_{i=1}^{d-1}\left(f_{x^{(i)},y^{(i)}}(n)\right) f_{x^{(d)},y^{(d)}}^{(+)}(n) \qquad (4.9)$$

$$= \prod_{i=1}^{d-1}\left(\widetilde{f}_{x^{(i)},y^{(i)}}(n)\right) \widetilde{f}_{x^{(d)},y^{(d)}}^{(+)}(n) \ . \qquad (4.10)$$

The first integral in the r.h.s of (4.8) causes no trouble, cf. Lemma 1.16 and 3.22. We turn to the second and third expression in (4.8).
Let us begin with the pinning case. With Lemma 1.16 we obtain for $n, m \geq 2$

$$\int_{\mathbb{R}^d} \widetilde{f}_{0,y}(n)\widetilde{f}_{0,y}(m)\,dy = \prod_{i=1}^{d}\left(\int_{\mathbb{R}} \widetilde{f}_{0,y^{(i)}}(n)\widetilde{f}_{0,y^{(i)}}(m)\,dy^{(i)}\right) \leq \text{const.}(nm)^{-d/2}\ ,$$

which is summable for $d \geq 3$ and

$$\int_{\mathbb{R}^d}\int_{\mathbb{R}^d} \widetilde{f}_{x,y}(n)\widetilde{f}_{x,y}(m)\,dx\,dy = \prod_{i=1}^{d}\left(\int_{\mathbb{R}}\int_{\mathbb{R}} \widetilde{f}_{x^{(i)},y^{(i)}}(n)\widetilde{f}_{x^{(i)},y^{(i)}}(m)\,dx^{(i)}\,dy^{(i)}\right) \leq \text{const.}(n+m)^{-d/2}\ .$$

But

$$\sum_{n,m=2}^{\infty}(n+m)^{-d/2} < \infty \iff d \geq 5$$

and this means B^0 is compact for $d \geq 5$ and so is B^θ on $L^2(\mathbb{R}^d, d\mu)$ for every $\theta \geq 0$.
We consider now the wetting case. By $\widetilde{f}_{x^{(i)},y^{(i)}}(n) \leq \text{const.}n^{-1/2}$ for $i = 1, ..., d-1$ and Lemma 3.22 we have for $n, m \geq N_0$ ($\mathbb{N}_{\geq 2}^2 \setminus \{(n,m)\,|\,n,m \geq N_0\}$ can be treated similarly)

$$\int_{\mathbb{R}^d} \widetilde{f}_{0,y}^{+}(n)\widetilde{f}_{0,y}^{+}(m)\,dy \leq \frac{\text{const.}}{(nm)^{(d-1)/2}}\frac{\log(n)\log(m)}{(nm)^{3/2}} = \text{const.}\frac{\log(n)\log(m)}{(nm)^{(d+2)/2}}$$

which is summable for $d \geq 1$ and furthermore for $d \geq 2$ we have

$$\int_{\mathbb{R}^d}\int_{\mathbb{R}^d} \widetilde{f}_{x,y}^{+}(n)\widetilde{f}_{x,y}^{+}(m)\,dx\,dy \leq \frac{\text{const.}}{(n+m)^{(d-1)/2}}\frac{\log(n)\log(m)}{(nm)^{3/2}}n^{1/2}$$

$$\leq \frac{\text{const.}}{(n+m)^{3/2}}\frac{\log(n)\log(m)}{n}\frac{1}{m^{3/2}} \leq \frac{\log(n)\log(m)}{n^{3/2}}\frac{\log(n)\log(m)}{m^{3/2}}\ .$$

So $B^{\theta,+}$ is compact on $L^2(\mathbb{R}^d, d\mu)$ for every $\theta \geq 0$ and all $d \geq 2$. The case $d = 1$ was treated in Lemma 3.22. □

4.4.2 Compactness for pinning in lower dimensions

Now we want to consider the lower dimensions. As was just seen, the Hilbert-Schmidt property fails in this case. Fortunately we will see that compactness can be established on an other space, which due to Zerner's Theorem, is still fine for furher investigations.

For this purpose let us recall a compactness criterion for integral operators on L^1, which is more than we need, cf [14].

Theorem 4.7 Let Ω be a measurable subset of $\mathbb{R}^d$ and $k : \Omega \times \Omega \to \mathbb{R}$ be a measurable function where there exists a constant $M > 0$ such that for almost all $x \in \Omega$, $k(x,.) \in L^1(\Omega)$ and $\int_\Omega |k(x,y)|\, dy < M$. Define operators T and T_* on $L^\infty(\Omega)$ and $L^1(\Omega)$ respectively by

$$(Tu)(x) = \int_\Omega k(x,y) u(y)\, dy ,$$

$$(T_*v)(y) = \int_\Omega k(x,y) v(x)\, dx ,$$

and define $\widetilde{k} : \Omega \times \mathbb{R}^d \to \mathbb{R}$ by

$$\widetilde{k}(x,y) = \begin{cases} k(x,y) & , \text{ if } y \in \Omega, \\ 0 & , \text{ if } y \in \mathbb{R}^d \setminus \Omega . \end{cases}$$

Then the following are equivalent:

(1) T is compact

(2) T_* is compact

(3) Given $\varepsilon > 0$ there exists $\delta > 0$ and $R > 0$ such that for almost all $x \in \Omega$ and for each $h \in \mathbb{R}^d$ with $|h| < \delta$,

$$\int_{\mathbb{R}^d \setminus B(0,R)} |\widetilde{k}(x,y)|\, dy < \varepsilon, \quad \int_{\mathbb{R}^d} |\widetilde{k}(x,y+h) - \widetilde{k}(x,y)|\, dy < \varepsilon . \qquad (4.11)$$

Now we can show the compactness in lower dimensions.

Proposition 4.8
For $d \in \{3,4\}$ and $\theta \geq 0$ the operator B^θ is compact on $L^1(\mathbb{R}^d, d\mu)$.

Proof To show this statement we will apply the last theorem. Take $\Omega = \mathbb{R}^d$ and $k(x,y) = b^\theta_{x,y}$, for an arbitrary but fixed $\theta \geq 0$. First we show that $b^\theta_{x,y}$ is uniformly bounded in $x \in \mathbb{R}$ on $L^1(\mathbb{R}^d, d\mu)$. To see this we use the representation (1.21) and the notation from Lemma 3.21 and compute (cf. formula A.11)

$$\int_{\mathbb{R}^d} \widetilde{f}_{x,y}(n)\, dy = \prod_{i=1}^d \left(\int_{\mathbb{R}} \widetilde{f}_{x^{(i)}, y^{(i)}}(n)\, dy^{(i)} \right) = \prod_{i=1}^d \left(\int_{\mathbb{R}} \varphi_n^{(-x^{(i)}, 0)}(y^{(i)}, 0)\, dy^{(i)} \right)$$

$$= \prod_{i=1}^d \left(\frac{1}{\sqrt{\det \Sigma_n} \sqrt{2\pi H_{1,1}(n)}} \exp\left\{ -\left(\frac{\det(\Sigma_n^{-1})}{2 H_{1,1}(n)} \left[\gamma \frac{1-\gamma^n}{1-\gamma} \right]^2 \left(x^{(i)} \right)^2 \right) \right\} \right) \qquad (4.12)$$

$$\leq \left(\frac{1}{\sqrt{2\pi H_{1,1}(n)}} \frac{1}{\sqrt{\det \Sigma_n}} \right)^d \leq \frac{\text{const.}}{n^{d/2}} ,$$

4.4 THE WEAK PINNING

where we recall $0 < H_{1,1}(n) \to$ const. , $n \to \infty$. With that we obtain finally the bound for some constant $0 < M < \infty$

$$\int_{\mathbb{R}^d} B^\theta_{x,dy} = \int_{\mathbb{R}^d} b^\theta_{x,y}\, \mu(dy) = \sum_{n \in \mathbb{N}} e^{-\theta n} \int_{\mathbb{R}^d} \widetilde{f}_{x,y}(n)\, \mu(dy) \qquad (4.13)$$

$$\leq \sum_{n \in \mathbb{N}} e^{-\theta n} \left(c \cdot n^{-d/2} + \widetilde{f}_{x,0}(n) \right)$$

$$\leq c \sum_{n \in \mathbb{N}} e^{-\theta n} n^{-d/2} \leq M \quad \text{for all } x \in \mathbb{R} \text{ and } \theta \geq 0,$$

where c denotes different constants and the last but one expression is of course summable for all $\theta \geq 0$, iff $d \geq 3$. Here we see that this argument fails for $\theta = 0$ and $d = 1$ or $d = 2$. What is left are the conditions in (4.11). They are quite technical to prove for our kernel $B^\theta_{x,dy}$. We start by proving the first one, i.e. with (4.13):

$$\lim_{R \to \infty} \sup_{x \in \mathbb{R}} \sum_{n \in \mathbb{N}} \int_{\mathbb{R}^d \setminus B(0,R)} \widetilde{f}_{x,y}(n)\, \mu(dy) = 0 \ . \qquad (4.14)$$

To make use of the product structure of $\widetilde{f}_{x,y}(n)$ we use $(-r,r)^d \subseteq B(0,R)$ where $r := R/\sqrt{2}$. In the last one of the following estimates we bound by (4.12)

$$\int_{\mathbb{R}^d \setminus B(0,R)} \widetilde{f}_{x,y}(n)\, dy \leq d(d-1) \int_{[-r,r]} \int_{\mathbb{R} \setminus B(0,r)} \int_{\mathbb{R}^{d-2}} \widetilde{f}_{x,y}(n)\, dy + d \int_{\mathbb{R} \setminus B(0,r)} \int_{\mathbb{R}^{d-1}} \widetilde{f}_{x,y}(n)\, dy$$

$$\leq 2d(d-1) \int_{\mathbb{R} \setminus B(0,r)} \int_{\mathbb{R}^{d-1}} \widetilde{f}_{x,y}(n)\, dy$$

$$\leq \frac{c_d}{n^{(d-1)/2}} \int_{\mathbb{R} \setminus B(0,r)} \widetilde{f}_{x^{(1)},y^{(1)}}(n)\, dy^{(1)} \ . \qquad (4.15)$$

Wlog we have chosen $y^{(1)}, y^{(2)}$ such that there is a maximum of the different permutations of the upper integrals in $x^{(1)}, x^{(2)}$. So we have reduced the problem to a one dimensional one. Here we can calculate similarly to (4.12), cf. formula A.11, that for $x \in \mathbb{R}$

$$\int_r^\infty \widetilde{f}_{x,y}(n)\, dy = \frac{1}{2\sqrt{\det \Sigma_n} \sqrt{2\pi H_{1,1}(n)}}$$

$$\cdot \exp\left\{ -\left(\frac{\det(\Sigma_n^{-1})}{2H_{1,1}(n)} \left[\gamma \frac{1-\gamma^n}{1-\gamma} \right]^2 r^2 \right) \right\} \left(1 + \mathrm{Erf}\left(\frac{-\frac{\gamma}{1-\gamma} c(\gamma,n) x - H_{1,1}(n) r}{\sqrt{2 H_{1,1}(n)}} \right) \right)$$

$$\qquad (4.16)$$

where again we used the notation from the proof of Lemma 3.21. Recall that for all γ, n it is const.$\geq H_{1,1}(n), c(\gamma,n) > 0$ and Erf denotes the "error function".

First we fix r and n. Then the supremum of (4.16) is attained in, say $x(r)$, which is finite, otherwise (4.16) would be zero and that is not possible for the maximum. Now we distinguish between two cases. Assume first that $x(r)$ is bounded in $r > 0$, then for r large enough the $\sup_{x \in \mathbb{R}}$ of (4.16) can be bounded from above by

$$\frac{1 + \mathrm{Erf}(-cr)}{2\sqrt{\det \Sigma_n} \sqrt{2\pi H_{1,1}(n)}} \longrightarrow 0 \quad , \text{ for } r \to \infty \ ,$$

for some constant $c > 0$. Otherwise if we assume that $|x(r)| \to \infty$ for $r \to \infty$, then the $\sup_{x \in \mathbb{R}}$ of (4.16) can be bounded from above by

$$\frac{1}{2\sqrt{\det \Sigma_n}\sqrt{2\pi H_{1,1}(n)}} \cdot \exp\left\{-\left(\frac{\det(\Sigma_n^{-1})}{2H_{1,1}(n)}\left[\gamma\frac{1-\gamma^n}{1-\gamma}\right]^2 x(r)^2\right)\right\} \longrightarrow 0 \quad, \text{ for } r \to \infty \,.$$

Therefore for fixed $n \in \mathbb{N}$ we have shown

$$\lim_{r \to \infty} \sup_{x \in \mathbb{R}} \int_r^\infty \widetilde{f}_{x,y}(n)\,dy = 0 \,.$$

In an analogous way we can prove the same for $\int_{-\infty}^{-r} \widetilde{f}_{x,y}(n)\,dy$, cf. calculation A.11. Thus with 4.15 we have shown that for a fixed $n \in \mathbb{N}$

$$\lim_{R \to \infty} \sup_{x \in \mathbb{R}} \int_{\mathbb{R}^d \setminus B(0,R)} \widetilde{f}_{x,y}(n)\,dy \leq \frac{c_d}{n^{(d-1)/2}} \lim_{R \to \infty} \sup_{x \in \mathbb{R}} \int_{\mathbb{R} \setminus B(0,r)} \widetilde{f}_{x^{(1)},y^{(1)}}(n)\,dy^{(1)} = 0 \,.$$

In other words, for an $\varepsilon > 0$ there exists an $R(\varepsilon, n)$, such that

$$\sup_{x \in \mathbb{R}} \int_{\mathbb{R}^d \setminus B(0,R(\varepsilon,n))} \widetilde{f}_{x,y}(n)\,dy < \frac{c_d}{n^{d/2}}\varepsilon \,, \tag{4.17}$$

with some constant $c_d > 0$ only dependent on the dimension d. If we can show that

$$\sup_{n \in \mathbb{N}} R(\varepsilon, n) < \infty \tag{4.18}$$

then indeed, for every $\varepsilon > 0$ there exists an $R(\varepsilon)$, such that

$$\sup_{x \in \mathbb{R}} \sum_{n \in \mathbb{N}} \int_{\mathbb{R}^d \setminus B(0,R(\varepsilon))} \widetilde{f}_{x,y}(n)\,\mu(dy) \leq \varepsilon \sum_{n \in \mathbb{N}} \frac{c_d}{n^{d/2}} \,. \tag{4.19}$$

This means we would have proven (4.14) and so the first condition in (4.11).
So, let us prove (4.18). For this purpose we write with the help of the proof of Lemma 3.21

$$\sup_{x \in \mathbb{R}} \sum_{n \in \mathbb{N}} \int_{\mathbb{R}^d \setminus B(0,R)} \widetilde{f}_{x,y}(n)\,\mu(dy) \leq \sum_{n \in \mathbb{N}} \int_{\mathbb{R}^d \setminus B(0,R)} \sup_{x \in \mathbb{R}} \widetilde{f}_{x,y}(n)\,\mu(dy)$$

$$= \sum_{n \in \mathbb{N}} \left(\frac{1}{2\pi\sqrt{\det(\Sigma_n)}}\right)^d \int_{\mathbb{R}^d \setminus B(0,R)} \sup_{x \in \mathbb{R}} e^{-\frac{1}{2}\left(\|y\|_2^2 a(\gamma,n) + \frac{\|x\|_2^2\gamma^2}{(1-\gamma)^2}b(\gamma,n) + \frac{2\langle x,y\rangle\gamma}{1-\gamma}c(\gamma,n)\right)}\,dy$$

$$\leq \sum_{n \in \mathbb{N}} \left(\frac{1}{2\pi\sqrt{\det(\Sigma_n)}}\right)^d \int_{\mathbb{R}^d \setminus B(0,R)} e^{-\frac{1}{2}\|y\|_2^2\left(a(\gamma,n) - \frac{c(\gamma,n)^2}{b(\gamma,n)}\right)}\,dy \,, \tag{4.20}$$

where the last step is due to the fact that the

$$\inf_{x \in \mathbb{R}} \left(\|y\|_2^2 a(\gamma,n) + \frac{\|x\|_2^2\gamma^2}{(1-\gamma)^2}b(\gamma,n) + \frac{2\langle x,y\rangle\gamma}{1-\gamma}c(\gamma,n)\right)$$

is attained in $x = -y\frac{(1-\gamma)c(\gamma,n)}{\gamma b(\gamma,n)}$, which is easily seen by computing the derivatives. Of course, if (.) in the exponent of (4.20) was greater than a positive constant for all $n \in \mathbb{N}$,

4.4 The weak pinning

then we were done. Unfortunately this is not the case, as numerical results show. Therefore the previous calculations to obtain (4.17) for fixed n were needed. Nevertheless for n large enough by the definition it can be seen that

$$a(\gamma,n) - \frac{c(\gamma,n)^2}{b(\gamma,n)} \underset{n\to\infty}{\sim} \frac{1-\gamma^2}{\sigma^2} - \frac{1}{n\sigma^2} \ .$$

Now for each $\alpha, \beta > 0$, and so γ, σ^2, we can choose an $m_{\alpha,\beta}$ such that

$$\frac{1-\gamma^2}{\sigma^2} - \frac{1}{n\sigma^2} \geq \frac{1-\gamma^2}{\sigma^2} - \frac{1}{m_{\alpha,\beta}\sigma^2} =: c_{\alpha,\beta} > 0 \quad , \text{for all } n \geq m_{\alpha,\beta} \ .$$

Therefore by (4.20) for every $\varepsilon > 0$ we can choose an $R(\varepsilon, m_{\alpha,\beta}) < \infty$ such that

$$\sup_{x\in\mathbb{R}} \sum_{n\geq m_{\alpha,\beta}} \int_{\mathbb{R}^d \setminus B(0,R(\varepsilon,m_{\alpha,\beta}))} \tilde{f}_{x,y}(n)\,\mu(dy) \leq \sum_{n\geq m_{\alpha,\beta}} \frac{c_d}{n^{d/2}} \int_{\mathbb{R}^d \setminus B(0,R(\varepsilon,m_{\alpha,\beta}))} e^{-\frac{1}{2}\|y\|_2^2 c_{\alpha,\beta}}\,dy$$

$$\leq \varepsilon \sum_{n\geq m_{\alpha,\beta}} \frac{c_d}{n^{d/2}}$$

Finally, if we take $R(\varepsilon) := \sup\{R(\varepsilon, n) \mid n = 1, ..., m_{\alpha,\beta}\}$ then the last inequality and (4.17) lead to (4.19) and this proves the first condition in (4.11).

For the second condition in (4.11) observe that

$$\int_{\mathbb{R}^d} |b^\theta_{x,y+h} - b^\theta_{x,y}|\,\mu(dy) = |b^\theta_{x,h} - b^\theta_{x,0}| + \int_{\mathbb{R}^d} |b^\theta_{x,y+h} - b^\theta_{x,y}|\,dy$$

$$\leq \sum_{n\in\mathbb{N}} e^{-\theta n} |\tilde{f}_{x,h}(n) - \tilde{f}_{x,0}(n)| + \sum_{n\in\mathbb{N}} e^{-\theta n} \int_{\mathbb{R}^d} |\tilde{f}_{x,y+h}(n) - \tilde{f}_{x,y}(n)|\,dy$$

$$=: \sum_{n\in\mathbb{N}} e^{-\theta n} \left[I_{x,h}(n) + II_{x,h}(n)\right] \ .$$

We are going to show that: $\forall \varepsilon > 0\ \exists \delta > 0\ \forall x \in \mathbb{R}^d\ \forall n \in \mathbb{N}\ \forall |h| < \delta$

$$I_{x,h}(n) < \frac{\varepsilon}{2} \frac{c_d}{n^{d/2}} \quad \text{and} \quad II_{x,h}(n) < \frac{\varepsilon}{2} \frac{c_d}{n^{d/2}} \ . \tag{4.21}$$

So altogether we will have: $\forall \tilde{\varepsilon} > 0\ \exists \delta > 0\ \forall x \in \mathbb{R}^d\ \forall \theta \geq 0\ \forall |h| < \delta$

$$\int_{\mathbb{R}^d} |b^\theta_{x,y+h} - b^\theta_{x,y}|\,\mu(dy) \leq \varepsilon \sum_{n\in\mathbb{N}} \frac{c_d}{n^{d/2}} \leq \tilde{\varepsilon} \ ,$$

where of course ε has been chosen as $\varepsilon = \tilde{\varepsilon}/\sum_{n\in\mathbb{N}} c_d\, n^{-d/2}$. Finally Theorem 4.7 could be applied, because its assumptions were fulfilled.

Let us start by showing the first part in (4.21). We write as before

$$I_{x,h}(n) = \left(\frac{1}{2\pi\sqrt{\det(\Sigma_n)}}\right)^d e^{-\frac{\|x\|_2^2 \gamma^2}{2(1-\gamma^2)} b(\gamma,n)} \left| e^{-\frac{1}{2}\left(\|h\|_2^2 a(\gamma,n) + \frac{2\langle x,h\rangle \gamma}{1-\gamma} c(\gamma,n)\right)} - 1 \right| \tag{4.22}$$

We are going to take the supremum in x and n, say $x(h), n(h)$, and then let $h \to 0$. Because of the complexity of this expression we will investigate all candidates and verify in all cases that $\sup_{x \in \mathbb{R}} \sup_{n \in \mathbb{N}} I_{x,h}(n) \to 0$ as $h \to 0$. Hereto we recall

$$a(\gamma, n) = O(1) \quad \text{and} \quad b(\gamma, n), c(\gamma, n) = O(1/n) \ .$$

Now let us fix an $h \in \mathbb{R}$, then there are the following possibilities for the supremum of (4.22).

- $||x(h)||_2 = \infty$ and $n(h) < \infty$. In this case $\lim_{||x||_2 \to \infty} I_{x,h}(n(h)) = 0$ and so it can't be the supremum.
- $||x(h)||_2 < \infty$ and $n(h) = \infty$. In this case

$$\lim_{n \to \infty} I_{x(h),h}(n) = \left(\frac{1}{2\pi \sqrt{\det(\Sigma_n)}} \right)^d \left| e^{-c||h||_2^2} - 1 \right| \ , \ c > 0$$

but this obviously tends to zero as $h \to 0$.
- $||x(h)||_2^2 = \infty$ and $n(h) = \infty$. In this case one has to distinguish between three cases.
  - $||x(h)||_2 = O(n(h))$. Here

$$\lim_{n, ||x||_2 \to \infty} I_{x,h}(n) = \left(\frac{1}{2\pi \sqrt{\det(\Sigma_n)}} \right)^d e^{-k/2} \left| e^{-c||h||_2^2} - 1 \right| \ , \ c, k > 0$$

  and again this tends to zero as $h \to 0$.
  - $||x(h)||_2^2 / n(h) \to \infty$. Here $\lim_{n, ||x||_2 \to \infty} I_{x,h}(n) = 0$, i.e. not a maximum.
  - $||x(h)||_2^2 = o(n(h))$. Here

$$\lim_{n, ||x||_2 \to \infty} I_{x,h}(n) = \left(\frac{1}{2\pi \sqrt{\det(\Sigma_n)}} \right)^d \left| e^{-c||h||_2^2} - 1 \right| \xrightarrow[h \to 0]{} 0 \ , \ c > 0 \ .$$

- $||x(h)||_2 < \infty$ and $n(h) < \infty$. In this case we distinguish between two cases
  - $\exists M > 0 : \sup\{x(h) \,|\, ||h||_2 \le M\} < \infty$. Here we have

$$I_{x(h),h}(n(h)) \le \left(\frac{1}{2\pi \sqrt{\det(\Sigma_n)}} \right)^d \left| e^{-\frac{1}{2} \left(||h||_2^2 a(\gamma, n(h)) + \frac{2\langle x(h), h \rangle \gamma}{1-\gamma} c(\gamma, n(h)) \right)} - 1 \right| \xrightarrow[h \to 0]{} 0 \ .$$

  - $||x(h)||_2 \to \infty$, if $h \to 0$. Here also $\lim_{h \to 0} I_{x(h),h}(n(h)) = 0$.

Therefore the first part in (4.21) has been shown. Now let us consider the second part. Choose an arbitrary $\varepsilon > 0$, then by the triangle-inequality and the previously proven result (4.14), for every $\delta > 0$ there exists an $R := R(\varepsilon, \delta) > 0$ such that for all $x \in \mathbb{R}^d, n \in \mathbb{N}$ and $|h| < \delta$

$$\int_{\mathbb{R}^d} |\tilde{f}_{x,y+h}(n) - \tilde{f}_{x,y}(n)| \, dy = \int_{\mathbb{R}^d \setminus B(0,R)} |\tilde{f}_{x,y+h}(n) - \tilde{f}_{x,y}(n)| \, dy + \int_{B(0,R)} |\tilde{f}_{x,y+h}(n) - \tilde{f}_{x,y}(n)| \, dy$$

$$\le \int_{\mathbb{R}^d \setminus B(0,R)} \tilde{f}_{x,y+h}(n) \, dy + \int_{\mathbb{R}^d \setminus B(0,R)} \tilde{f}_{x,y}(n) \, dy + \int_{B(0,R)} |\tilde{f}_{x,y+h}(n) - \tilde{f}_{x,y}(n)| \, dy$$

$$< \frac{2}{3}\varepsilon + \int_{B(0,R)} |\tilde{f}_{x,y+h}(n) - \tilde{f}_{x,y}(n)| \, dy \ ,$$

Observe that in order to do that we have enlarged R w.r.t. δ such that adding h to y makes no problem. Apparently $R(\varepsilon, \delta) \nrightarrow \infty$ when $\delta \to 0$, so we can choose now $R(\varepsilon)$ independently of δ. Therefore we can find a $\delta > 0$ such that for all $|h| < \delta$

$$\sup_{x \in \mathbb{R}^d} \sup_{n \in \mathbb{N}} \int_{B(0,R)} |\tilde{f}_{x,y+h}(n) - \tilde{f}_{x,y}(n)| \, dy \leq \text{Vol}(B(0,R)) \sup_{x \in \mathbb{R}^d} \sup_{n \in \mathbb{N}} \sup_{y \in B(0,R)} |\tilde{f}_{x,y+h}(n) - \tilde{f}_{x,y}(n)|$$
$$< \frac{\varepsilon}{3}.$$

The proof of the last step works in the same way as the proof for the first part in (4.21), since y is bounded by the ball $B(0, R)$. Now we have finished, because we proved (4.21) and the lines thereafter conclude this Proposition. □

Remark 4.9
The difference to the dimensions $d = 1$ and $d = 2$ can be seen by (4.12). Namely, we know that
$$\det(\Sigma_n^{-1}) = O(1/n).$$
This means by (4.13) that for each $x \in \mathbb{R}^d$
$$(B^\theta \text{Id})(x) = \int_{\mathbb{R}^d} B^\theta_{x,dy} \nearrow \infty, \text{ for } \theta \searrow 0.$$
This indicates that for $d = 1, 2$ we have a completely different behavior. Indeed, in the spirit of the proof of Proposition 1.17 we obtain
$$\lim_{\theta \searrow 0} \left(B^\theta\right)^{\circ 2}_{0,\{0\}} \geq \text{const.} \sum_{n=2}^{\infty} \frac{1}{n^{d/2}},$$
which is not summable for $d = 1, 2$. This means by the variational formula (1.33) for the spectral radius δ_θ of B^θ that we have $\delta_\theta \nearrow \infty$, for $\theta \searrow 0$.

4.4.3 Conclusion and results

Remark 4.10
For the pinning model in dimensions $d = 3$ and $d = 4$ it can be shown that B^θ is also compact on $L^2(\mathbb{R}^d, d\mu)$ for $\theta > 0$. For $\theta = 0$ we have just seen the compactness of B^0 only on $L^1(\mathbb{R}^d, d\mu)$. Nevertheless this is still good enough and the infinite-dimensional Perron-Frobenius Theorem of Zerner A.1 still applies. As a consequence the Markov chain $\{J_k\}_k$ is positive recurrent for $\varepsilon \geq \varepsilon_c$. This can be seen similarly to Remark 1.18 and 3.23. The difference here is that the right eigenvalue of $B^{F(\varepsilon)}$ is located in $L^1(\mathbb{R}, d\mu)$ and the left one in $L^\infty(\mathbb{R}, d\mu)$. Thus we can state that the process $(\eta_i)_{i \in \mathbb{Z}^+}$ under $\mathcal{P}_\varepsilon$ is a classical renewal process, which is non-terminating if $\varepsilon \geq \varepsilon_c$.

These were the crucial steps comparing to the $(1+1)$-dimensional cases. The identification of the free energy is done in the same way, for instance we obtain similar formula to (1.31), i.e.
$$\mathcal{Z}^{(+)}_{\varepsilon,N,d,0} = \frac{e^{F_s^{(+),d}(\varepsilon)(N+1)}}{\varepsilon^2} \lambda^{d(N+1)} \mathcal{P}^{(+),d}_\varepsilon(\mathcal{A}_N).$$

Then one has to verify that $F_s^{(+),d}$ indeed equals to the free energy, but verifying this does not depend on the dimension. Like mentioned in the beginning, we refer to chapters 1 and 3. Finally one arrives at

Theorem 4.11 (Mixed pinning model)
For the weak pinning in the (pure) pinning model we have a trivial phase transition in dimensions $d = 1, 2$. Whereas for $d \geq 3$ there is a non-trivial phase transition.

and

Theorem 4.12 (Mixed wetting model)
For the weak pinning in the wetting model we have in every dimension $d \geq 1$ a non-trivial phase transition. Moreover $0 \leq \varepsilon_{c,d} < \varepsilon_{c,d}^+ < \infty$ and $\varepsilon_{c,d}^{(+)}$ is non-decreasing in d.

Remark 4.13 *Showing that $0 \leq \varepsilon_{c,d} < \varepsilon_{c,d}^+ < \infty$ is done in the same way as in [10], where an perturbation result of [21] has been used. There one only needed that $B^{0,d}$ and $B^{0,+,d}$ are compact on $L^2(R^d, d\mu)$, but this is now clear for $d \geq 5$. For $d = 3, 4$ the same argument can be used also with compactness on $L^1(R^d, d\mu)$. The fact that $\varepsilon_{c,d}^{(+)}$ is non-decreasing in d can be seen the same way like in (4.23) and the line after.*

4.4.4 Remark on the Laplacian model

In the case of pure models, i.e. where in (4.2) only one $q_j \neq 0$, one can obtain much more easier explicit expressions. For instance for the Gaussian Laplacian model, i.e. $\alpha = 0$, in the $(1+1)$-dimensional case we computed the following expressions:

$$\det \Sigma_n = \frac{\det(\beta B_{n-2})}{\beta^n} = \frac{n^2(n^2-1)}{12\beta^2} \quad , \lambda = \left(\frac{2\pi}{\beta}\right)^{1/2} \quad , \nu(x) = 1 \,,$$

where

$$\Sigma_n = \frac{1}{6\beta}\begin{pmatrix} n(2n^2 - 3n + 1) & 2n(n^2 - 1) \\ 2n(n^2 - 1) & n(2n^2 + 3n + 1) \end{pmatrix}.$$

This means for $v, w \in \mathbb{R}$

$$\widetilde{f}_{v,w}(n) = \frac{1}{2\pi\sqrt{\det(\Sigma_n)}} \exp\left\{-\frac{1}{2}\left\langle \begin{pmatrix} w + (n-1)v \\ nv \end{pmatrix}, \Sigma_n^{-1}\begin{pmatrix} w + (n-1)v \\ nv \end{pmatrix}\right\rangle\right\} \mathbb{1}_{\{y \neq 0\}} \,.$$

Then it is easily seen that for $x, y \in \mathbb{R}^d$

$$\widetilde{f}_{x,y}(n) \leq \widetilde{f}_{x^{(i)}, y^{(i)}}(n) \quad , \text{ for all } i = 1, ...d \,.$$

Therefore also

$$B_{x,dy}^{\theta,(+),d} \geq B_{x,dy}^{\theta,(+),d+1} \,, \tag{4.23}$$

which in particular means that $\delta_d^{(+)}(0) \geq \delta_{d+1}^{(+)}(0)$ and so $\varepsilon_{c,d}^{(+)} \leq \varepsilon_{c,d+1}^{(+)}$. Here d denotes the corresponding expressions in the d-dimensional model. Now since the Gaussian case is covered by the assumptions on the potential in [10], it is clear that also

Conclusion 4.14
The Gaussian Laplacian model exhibits in all dimensions $d \geq 1$ a non-trivial phase transition, i.e.
$$0 < \varepsilon_c < \varepsilon_c^+ < \infty$$
and $\varepsilon_c^{(+)}$ is non-decreasing in d.

Although we have first studied the higher dimensions, of course the same argument applies to our mixed model, meaning that a "proper" phase transition in lower dimensions implies a "proper" phase transitions in higher dimensions.

Remark 4.15
At this point we conjecture that based on Remark 4.2 the only (Gaussian) models of type (4.1) which exhibit a trivial phase transition are those pinning-models (no wall) where $\inf\{i \geq 1 | q_i \neq 0\} = 1$ and the dimension d equals either one or two.

4.5 General pinning subspace M

We would like to generalize the heavy and weak cases to a general subspace M. Fix for this purpose an $r \in \{1, ...d\}$. Choose pairwise disjoint elements $i_j \in \{1, ..., d\}$ for $j = 1, ..., r$ and consider
$$M_{i_1,...,i_r} := \{x \in \mathbb{R}^d \, | \, x_{i_1} = \cdots = x_{i_r} = 0\} \, .$$
Because of the initial assumption on M, one should require in the wetting case one i_j to be equal to d. Without loss of generality we can from now on consider
$$M_m = \{x \in \mathbb{R}^d \, | \, x_d = x_{d-1} = \cdots = x_{d-(r-1)} = 0\} \, ,$$
where by m we have denoted the dimension of M. Here the product-form of the pinning measure can be written similarly to (4.4) as
$$\prod_{i=1}^{N-1} (\varepsilon \nu(d\varphi_i^{(+)}) + d\varphi_i^{(+)}) = \prod_{i=1}^{N-1} \left(\varepsilon d\varphi_i^{(1,...,d-r)} \delta_0(d\varphi_i^{(d-r+1,...,d)}) + d\varphi_i \right)$$
$$= \left(\coprod_{i=1}^{N-1} d\varphi_i^{(1)} \cdots d\varphi_i^{(d-r)} \right) \left(\coprod_{i=1}^{N-1} (\varepsilon \delta_0(d\varphi_i^{(d-r+1,...,d)}) + d\varphi_i^{(d-r+1,...,d)}) \right) \, .$$

Observing that $m = d - r$, we have therefore the following representation for the partition function
$$\mathcal{Z}_{\varepsilon,N,d,m}^{(+)} = (\mathcal{Z}_{0,N,1,0})^m \, \mathcal{Z}_{\varepsilon,N,d-m,0}^{(+)} \, , \qquad (4.24)$$

At this point we see that this model separates into a free model of dimension $dim(M) = m$, because
$$(\mathcal{Z}_{0,N,1,0})^m = \mathcal{Z}_{0,N,m,0}$$
and a weak pinning model of dimension $d - dim(M) = d - m$.

With the separation in (4.24) we are able to write the free energy as

$$F_{d,m}^{(+)}(\varepsilon) = \lim_{N\to\infty} \frac{1}{N} \log \left(\frac{\mathcal{Z}_{\varepsilon,N,d,m}^{(+)}}{\mathcal{Z}_{0,N,d,m}^{(+)}} \right) = \lim_{N\to\infty} \frac{1}{N} \log \left(\frac{(\mathcal{Z}_{0,N,1,0})^m \, \mathcal{Z}_{\varepsilon,N,d-m,0}^{(+)}}{(\mathcal{Z}_{0,N,1,0})^m \, \mathcal{Z}_{0,N,d-m,0}^{(+)}} \right) = F_{d-m,0}^{(+)}(\varepsilon) \, .$$

It is now evident that substantial for localization is the difference between the dimension of the state space of φ_i and $\dim(M)$, as we have already mentioned before. In other words:

Corollary 4.16
A model with a general pinning subspace M of dimension m behaves just like the corresponding weak pinning model with a $(d-m)$-dimensional state space.

Therefore immediately from the weak pinning investigation we conclude

Corollary 4.17
Let M be an m-dimensional pinning subspace with $0 \leq m \leq d-1$. Set $r := d-m$, then for the mixed pinning model we have a trivial phase transition if $r = 1$ or $r = 2$. The pinning and wetting models exhibit a non-trivial phase transition, if

(a) $r \geq 3$ in the pinning case

(b) $r \geq 1$ in the wetting case .

Furthermore we have in every dimension $d \geq 1$ the relation

$$0 \leq \varepsilon_c < \varepsilon_c^+ < \infty \, .$$

Moreover, in the localized regime the free energy is real analytic and behaves asymptotically like $\log \varepsilon$ for $\varepsilon \to \infty$.

Remark 4.18
Considering subsection 4.4.4, we can state the same as in Corollary 4.17 for the Laplacian model. Here we have a non-trivial phase transition if $r \geq 1$, for both, pinning and wetting.

5 Order of Phase Transitions for Gaussian models

This chapter is devoted to the study of localization and delocalization in terms of path properties of the linear chain in the critical regime. It was shown already in chapter 0 that in the supercritical regime ($\varepsilon > \varepsilon_c^{(+)}$) typically the average number of contacts to the x-axis behaves like $O(N)$, recall

$$\ell_N = \#\{k \in \{1,...,N\} \,|\, \varphi_k = 0\} \,. \tag{5.1}$$

Whereas in the subcritical regime the asymptotic is less than linear: $o(N)$. The critical regime requires a more careful study and is connected to the question of differentiability of the free energy.

5.1 Preliminaries and results

In this chapter we consider the model 4.1 (case (iii)) with weak pinning, i.e. the pinning space equals $M = \{0\}$. This is not a restriction as was already shown in section 4.5. We know already that the free energy is differentiable in the sub- and supercritical regime, since it is constant or analytic, respectively. But what happens at the criticality? One option is $(F_d^{(+)})'\left(\varepsilon_c^{(+)}\right) = 0$, that means a phase transition of higher order than one. Therefore, recall chapter 0, at the criticality we would have on average $o(N)$ contacts, so a delocalized behavior in the paths-sense. The other case occurs when the right derivative is strictly positive. Then the phase transition is said to be of first order and localization takes place, i.e. the typical paths touch the defect line $O(N)$-times. Recall that in any case it is always $F^{(+)}(\varepsilon_0^{(+)}) = 0$, therefore at the criticality we have always to distinguish in which sense we are speaking of localization/delocalization. We will prove the following results.

Theorem 5.1 (First order transition for the mixed model)
The weak pinning model displays a first order phase transition if $d \geq 5$, i.e. the right-derivative is strictly positive:

$$\lim_{\varepsilon \searrow \varepsilon_c^{(+)}} \frac{F^{(+)}(\varepsilon)}{\varepsilon - \varepsilon_c^{(+)}} > 0 \,.$$

Moreover, there is a first order phase transition for the (weak) wetting model already for $d \geq 3$.

Theorem 5.2 (Higher order transition for the mixed model)
We have a second order phase transition in the weak models if

(1) $d = 3, 4$ for the pinning-model,

(2) $d = 1$ for the pinning-model under assumption 5.14 and

(3) $d = 1, 2$ for the wetting-model under assumption **(CW)** in section 5.3.

Furthermore under assumption 5.14 we have an infinite order of phase transition for the pinning model if $d = 2$.

This means, in higher dimensions the smoothness of the phase transitions gets lost and the contact fraction has a jump at the critical point. In particular, for our model we can distinguish between three different states ordered by increasing dimension d, namely:

(i) trivial phase transition, which is smooth

(ii) (proper) phase transition, which is smooth

(iii) (proper) phase transition, which is of first order .

In case of the wetting model the state (i) is not existent.

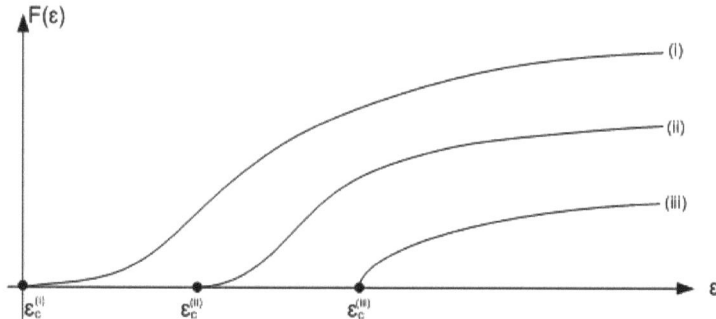

Figure 5.1: A sketch of the free energies in context of the upper three cases (i)-(iii).

Let us compare our results on the order of phase transition to the ones of the gradient and Laplacian model. Bolthausen, Funaki and Otobe have obtained in [7] the same behavior for the pure gradient case with Gaussian interaction potential. In this context it is remarkable that Caravenna and Deuschel obtained in [10] a first order phase transition for the $(1+1)$-dimensional wetting model with general potential. This punctuates again the crucial impact of the ∇-interaction in the mixed model. Moreover, they have also proven a second order transition for the $(1 + 1)$-dimensional pinning model.

5.2 First order phase transition

We will first prove the Theorem 5.1. In this case the renewal structure of our double-contact process $\{\eta_k\}_{k\geq 0}$ (cf. (1.37)) and the finiteness of the first moment of η_1 will be sufficient to prove the first order phase transition. To stay short in notation, we neglect the dimension parameter d in various expressions, but in this section we mean $d \geq 5$ for the pinning and $d \geq 3$ for the wetting model, unless otherwise stated.

Let us start with a Lemma about the contact number on the defect line.

Lemma 5.3 *For every $\varepsilon > 0$*

$$\mathbb{E}_{\mathbb{P}^{(+)}_{\varepsilon,N,d}}[\ell_N - 1] = \varepsilon \left(\frac{\partial}{\partial \varepsilon} \log \mathcal{Z}^{(+)}_{\varepsilon,N,d} \right) .$$

Proof Consider the expansion of the product measure in (1.22), set $M = \{1, ..., N-1\}$ and denote

$$R^{(+)}_{k,N} := \sum_{\substack{A \subseteq M \\ A = k}} \int_{(\mathbb{R}^d)^{N-1}} e^{-\mathcal{H}_{[-1,N+1]}(\varphi)} \left(\prod_{i \in A} \delta_0(d\varphi_i) \right) \left(\prod_{j \in M \setminus A} d\varphi_j^{(+)} \right) .$$

Now the verification of the Lemma is straight forward by calculation:

$$\varepsilon \left(\frac{\partial}{\partial \varepsilon} \log \mathcal{Z}^{(+)}_{\varepsilon,N,d} \right) = \frac{\varepsilon}{\mathcal{Z}^{(+)}_{\varepsilon,N,d}} \frac{\partial}{\partial \varepsilon} \mathcal{Z}^{(+)}_{\varepsilon,N,d} = \frac{\varepsilon}{\mathcal{Z}^{(+)}_{\varepsilon,N,d}} \sum_{k=0}^{N-1} k \varepsilon^{k-1} R^{(+)}_{k,N}$$

$$= \sum_{k=0}^{N-1} k \, \mathbb{P}^{(+)}_{\varepsilon,N,d}(\ell_N = k+1) = -1 + \sum_{k=0}^{N-1} (k+1) \, \mathbb{P}^{(+)}_{\varepsilon,N,d}(\ell_N = k+1)$$

$$= -1 + \mathbb{E}_{\mathbb{P}^{(+)}_{\varepsilon,N,d}}[\ell_N] .$$

$\square$

Next we will need an integrability condition for the eigenvalues of the operator $B^{F(\varepsilon),(+)}$, which was defined in (4.6).

Lemma 5.4 *For each $\varepsilon > 0$, the right- and left eigenvalues $\nu^{(+)}_\varepsilon$ and $w^{(+)}_\varepsilon$, corresponding to the spectral radius of $B^{F(\varepsilon),(+)}$, are bounded in L^1, i.e.*

$$\|\nu^{(+)}_\varepsilon\|_{L^1(\mathbb{R}^d,d\mu)} < \infty \quad \text{and} \quad \|w^{(+)}_\varepsilon\|_{L^1(\mathbb{R}^d,d\mu)} < \infty ,$$

if

- *$d \geq 5$ in pinning case and*
- *$d \geq 1$ in wetting case .*

Proof The proofs for the right- and left eigenvalues are analogous, so let us consider the one for $\nu_\varepsilon^{(+)}$. Due to (1.36) and (3.39) we have

$$\nu_\varepsilon^{(+)}(x) \leq c^{(+)}(\varepsilon) \sum_{n=1}^{\infty} \int_{\mathbb{R}^d} \widetilde{f}_{x,y}^{(+)}(n) \, \nu_\varepsilon^{(+)}(y) \, \mu(dy) \, , \tag{5.2}$$

where

$$c^{(+)}(\varepsilon) = \begin{cases} \varepsilon & \text{, in pinning case} \\ \max\{\varepsilon, \varepsilon_c^+\} & \text{, in wetting case} \end{cases}.$$

Therefore by the Cauchy-Schwarz inequality

$$\int_{\mathbb{R}^d} \nu_\varepsilon^{(+)}(x) \, \mu(dx) \leq c^{(+)}(\varepsilon) \|\nu_\varepsilon^{(+)}\|_{L^2(\mathbb{R}^d, d\mu)} \sum_{n=1}^{\infty} \left\| \int_{\mathbb{R}^d} \widetilde{f}_{x,\cdot}^{(+)}(n) \, \mu(dx) \right\|_{L^2(\mathbb{R}^d, d\mu)} \tag{5.3}$$

and this is of no harm, since by the last chapter we know that $\nu_\varepsilon^{(+)}, w_\varepsilon^{(+)} \in L^2(\mathbb{R}^d, d\mu)$. We first start by considering the pinning case. By Jensen and $\widetilde{f}_{x,y}(n) \leq (\text{const.}) n^{-d/2}$

$$\left\| \int_{\mathbb{R}^d} \widetilde{f}_{x,\cdot}(n) \, \mu(dx) \right\|_{L^2(\mathbb{R}^d, d\mu)}^2 = \int_{\mathbb{R}^d} \left(\int_{\mathbb{R}^d} \widetilde{f}_{x,y}(n) \, \mu(dx) \right)^2 \mu(dy) \leq \int_{\mathbb{R}^d} \int_{\mathbb{R}^d} \widetilde{f}_{x,y}^2(n) \, \mu(dx) \, \mu(dy)$$

$$\leq \frac{\text{const.}}{n^{d/2}} \left(\int_{\mathbb{R}^d} \int_{\mathbb{R}^d} \widetilde{f}_{x,y}(n) \, dx \, dy + \int_{\mathbb{R}^d} \widetilde{f}_{0,y}(n) \, dy + \int_{\mathbb{R}^d} \widetilde{f}_{x,0}(n) \, dx + \widetilde{f}_{0,0}(n) \right)$$

$$\leq \frac{\text{const.}}{n^{d/2}} \, ,$$

where the last inequality is due to the proof of Lemma 1.16 combined with Proposition 4.5. Therefore by (5.3) we obtain in case of $d \geq 5$ for some constant $c > 0$

$$\|\nu_\varepsilon\|_{L^1(\mathbb{R}^d, d\mu)} \leq c \|\nu_\varepsilon\|_{L^2(\mathbb{R}^d, d\mu)} \sum_{n=1}^{\infty} \frac{1}{n^{d/4}} < \infty \, .$$

If we wouldn't take out the factor $\widetilde{f}_{x,y}(n)$ out of the integration, we could even show that this is true for $d \geq 3$, but this will not help us in further investigation. Of course, since $\widetilde{f}_{x,y}^+(n) \leq \widetilde{f}_{x,y}(n)$, this is also true for the wetting and $d \geq 5$. However in the wetting case we can do better. Taking a look at the proof of Proposition 4.6 we can see that

$$\left\| \int_{\mathbb{R}^d} \widetilde{f}_{x,\cdot}^+(n) \, \mu(dx) \right\|_{L^2(\mathbb{R}^d, d\mu)}^2 = \int_{\mathbb{R}^d} \left(\int_{\mathbb{R}^d} \widetilde{f}_{x,y}^+(n) \, \mu(dx) \right)^2 \mu(dy)$$

$$\leq \int_{\mathbb{R}^d} \int_{\mathbb{R}^d} \left(\widetilde{f}_{x,y}^+(n) \right)^2 \mu(dx) \, \mu(dy)$$

$$= \left(\int_{\mathbb{R}^d} \int_{\mathbb{R}^d} \left(\widetilde{f}_{x,y}^+(n) \right)^2 dx \, dy + \int_{\mathbb{R}^d} \left(\widetilde{f}_{0,y}^+(n) \right)^2 dy + \int_{\mathbb{R}^d} \left(\widetilde{f}_{x,0}^+(n) \right)^2 dx + \left(\widetilde{f}_{0,0}^+(n) \right)^2 \right)$$

$$\leq \text{const.} \frac{(\log n)^2}{n^{(d+4)/2}} \, .$$

Again by (5.3) we have for $d \geq 1$ and some constant $c > 0$

$$\|\nu_\varepsilon^+\|_{L^1(\mathbb{R}^d, d\mu)} \leq c \|\nu_\varepsilon^+\|_{L^2(\mathbb{R}^d, d\mu)} \sum_{n=1}^{\infty} \frac{\log n}{n^{(d+4)/4}} < \infty \, .$$

5.2 First order phase transition

□

Following similarly to [10], we are interested in the first moment of η_1. Observe that for the first double-contact of the field we have $\varphi_{\eta_1} = \varphi_{\eta_1-1} = 0$. Since ζ_1 (cf. (1.38)) is the first time that $\{J_k\}_k$ hits zero, i.e. $J_{\zeta_1} = 0$, from the definition we have $J_{\zeta_1} = \varphi_{\tau_{\zeta_1}-1} = 0$ and so $\varphi_{\tau_{\zeta_1}} = 0$. Therefore the equality $\eta_1 = \tau_{\zeta_1}$ holds.

We define $\check{K}_{x,dy}^{(+),\varepsilon_c}(n) := K_{x,dy}^{(+),\varepsilon_c}(n)\mathbb{1}_{\{y\neq 0\}}$, the transition probability of the Markov chain $\{(\tau_k, J_k)\}_k$ before the chain $\{J_k\}_k$ comes back to zero. Recall that " $*$ " denotes the convolution of operators defined in section 0.4. Furthermore set $K_{x,dy}^{(+),\varepsilon_c}(0) := 0$. Now we have for $n \geq 2$

$$\mathcal{P}_{\varepsilon_c}^{(+)}(\eta_1 = n) = \mathcal{P}_{\varepsilon_c}^{(+)}(\tau_{\zeta_1} = n) = \sum_{k=1}^{\infty} \mathcal{P}_{\varepsilon_c}^{(+)}(\tau_k = n, \zeta_1 = k)$$

$$= \sum_{k=1}^{\infty} \mathcal{P}_{\varepsilon_c}^{(+)}\left((\tau_k, J_k) \in (\{n\}, \{0\}) \mid (\tau_0, J_0) = (0,0),\, J_1 \neq 0, ..., J_{k-1} \neq 0\right)$$

$$= \int_{y \in \mathbb{R}} \left(\sum_{k=1}^{\infty} \left(\check{K}^{(+),\varepsilon_c}\right)_{0,dy}^{*k}(n-1)\right) \cdot K_{y,\{0\}}^{(+),\varepsilon_c}(1) \quad (5.4)$$

and $\mathcal{P}_{\varepsilon_c}^{(+)}(\eta_1 = 1) = K_{0,\{0\}}^{(+),\varepsilon_c}(1)$. A convenient representation for the convolution term and $r \in \mathbb{N}, k \geq 2$ can be obtained by

$$\left(\check{K}^{(+),\varepsilon_c}\right)_{0,dy}^{*k}(r) = \sum_{r_1=0}^{r} \left(\check{K}^{(+),\varepsilon_c}(r_1) \circ \left(\check{K}^{(+),\varepsilon_c}\right)^{*(k-1)}(r-r_1)\right)_{0,dy}$$

$$= \sum_{r_1=0}^{r} \left(\check{K}^{(+),\varepsilon_c}(r_1) \circ \left(\sum_{r_2=0}^{r-r_1} \check{K}^{(+),\varepsilon_c}(r_2) \circ \left(\check{K}^{(+),\varepsilon_c}\right)^{*(k-2)}(r-r_1-r_2)\right)\right)_{0,dy}$$

$$= \cdots$$

$$= \sum_{r_1=0}^{r} \sum_{r_2=0}^{r-r_1} \cdots \sum_{r_{k-1}=0}^{r-r_1-\cdots-r_{k-2}} \left(\check{K}^{(+),\varepsilon_c}(r_1) \circ \cdots \circ \check{K}^{(+),\varepsilon_c}(r_{k-1}) \circ \check{K}^{(+),\varepsilon_c}(\underbrace{r-r_1-\cdots-r_{k-1}}_{=:r_k})\right)_{0,dy}$$

$$= \sum_{\substack{r_1,...,r_k \in \mathbb{N}_0 \\ r_1+\cdots+r_k=r}} \left(\check{K}^{(+),\varepsilon_c}(r_1) \circ \cdots \circ \check{K}^{(+),\varepsilon_c}(r_k)\right)_{0,dy}. \quad (5.5)$$

Recall from the pinning and wetting model the transition kernel $D_{x,dy}^{(+),\varepsilon_c}$ of the Markov chain $\{J_k\}_k$ and set similarly

$$\check{D}_{x,dy}^{(+),\varepsilon_c} := \sum_{n \in \mathbb{N}} \check{K}_{x,dy}^{(+),\varepsilon_c}(n) = D_{x,dy}^{(+),\varepsilon_c}\mathbb{1}_{\{y \neq 0\}} \quad \text{and} \quad \check{G}_{x,dy}^{(+),\varepsilon_c} := \sum_{n \in \mathbb{N}} n\check{K}_{x,dy}^{(+),\varepsilon_c}(n).$$

Before proving the first result on the order of the phase transition, we will need a small

Lemma 5.5 *For every $k \in \mathbb{N}$ and $i = 1, ..., k$ it holds*

$$\sum_{r=1}^{\infty} \sum_{\substack{r_1,...,r_k \in \mathbb{N}_0 \\ r_1+\cdots+r_k=r}} r_i \left(\check{K}^{(+),\varepsilon_c}(r_1) \circ \cdots \check{K}^{(+),\varepsilon_c}(r_k)\right)_{0,dy} = \left(\left(\check{D}^{(+),\varepsilon_c}\right)^{\circ(i-1)} \circ \check{G}^{(+),\varepsilon_c} \circ \left(\check{D}^{(+),\varepsilon_c}\right)^{\circ(k-i)}\right)_{0,dy}$$

Proof
The proof can be conducted by induction over k. For $k=1$ the equality is obvious. Now let us prove the statement for $k+1$

$$\sum_{r=1}^{\infty} \sum_{\substack{r_1,\ldots,r_{k+1}\in\mathbb{N}_0 \\ r_1+\cdots+r_{k+1}=r}} r_i \left(\check{K}^{(+),\varepsilon_c}(r_1) \circ \cdots \check{K}^{(+),\varepsilon_c}(r_{k+1})\right)_{0,dy}$$

$$= \sum_{r=1}^{\infty} \sum_{r_{k+1}=0}^{r} \sum_{r_1+\cdots+r_k=r-r_{k+1}} r_i \left(\check{K}^{(+),\varepsilon_c}(r_1) \circ \cdots \check{K}^{(+),\varepsilon_c}(r_{k+1})\right)_{0,dy}$$

$$= \sum_{r=1}^{\infty} \sum_{r_1+\cdots+r_k=r} r_i \left(\check{K}^{(+),\varepsilon_c}(r_1) \circ \cdots \check{K}^{(+),\varepsilon_c}(r_{k+1})\right)_{0,dy}$$

$$+ \sum_{r=1}^{\infty} \sum_{r_1+\cdots+r_k=r-1} r_i \left(\check{K}^{(+),\varepsilon_c}(r_1) \circ \cdots \check{K}^{(+),\varepsilon_c}(r_{k+1})\right)_{0,dy}$$

$$+ \sum_{r=2}^{\infty} \sum_{r_1+\cdots+r_k=r-2} r_i \left(\check{K}^{(+),\varepsilon_c}(r_1) \circ \cdots \check{K}^{(+),\varepsilon_c}(r_{k+1})\right)_{0,dy} + \cdots$$

$$= \sum_{r_{k+1}=0}^{\infty} \sum_{r=r_{k+1}}^{\infty} \sum_{r_1+\cdots+r_k=r-r_{k+1}} r_i \left(\check{K}^{(+),\varepsilon_c}(r_1) \circ \cdots \check{K}^{(+),\varepsilon_c}(r_{k+1})\right)_{0,dy}$$

$$= \sum_{r_{k+1}=0}^{\infty} \sum_{r=0}^{\infty} \sum_{r_1+\cdots+r_k=r} r_i \left(\check{K}^{(+),\varepsilon_c}(r_1) \circ \cdots \check{K}^{(+),\varepsilon_c}(r_{k+1})\right)_{0,dy}$$

$$= \int_{z\in\mathbb{R}^d} \sum_{r=0}^{\infty} \sum_{r_1+\cdots+r_k=r} r_i \left(\check{K}^{(+),\varepsilon_c}(r_1) \circ \cdots \check{K}^{(+),\varepsilon_c}(r_k)\right)_{0,dz} \cdot \sum_{r_{k+1}=0}^{\infty} \check{K}^{(+),\varepsilon_c}_{z,dy}(r_{k+1})$$

$$= \int_{z\in\mathbb{R}^d} \left(\left(\check{D}^{(+),\varepsilon_c}\right)^{\circ(i-1)} \circ \check{G}^{(+),\varepsilon_c} \circ \left(\check{D}^{(+),\varepsilon_c}\right)^{\circ(k-i)}\right)_{0,dz} \cdot \left(\check{D}^{(+),\varepsilon_c}\right)_{z,dy}$$

$$= \left(\left(\check{D}^{(+),\varepsilon_c}\right)^{\circ(i-1)} \circ \check{G}^{(+),\varepsilon_c} \circ \left(\check{D}^{(+),\varepsilon_c}\right)^{\circ(k+1-i)}\right)_{0,dy}.$$

$\square$

Proof of Theorem 5.1
Let us now write the first moment of η_1 with the help of the calculations above. Here we

5.2 First order phase transition

use (5.4), (5.5) and Lemma 5.5 applied k-times with $n = r_1 + \cdots + r_k$ below

$$\mathbb{E}_{\mathcal{P}_{\varepsilon_c}^{(+)}}[\eta_1] = \sum_{n \in \mathbb{N}} n \mathcal{P}_{\varepsilon_c}^{(+)}(\eta_1 = n) = 1 + \sum_{n=1}^{\infty} n \mathcal{P}_{\varepsilon_c}^{(+)}(\eta_1 = n + 1)$$

$$= 1 + \sum_{n=1}^{\infty} n \int_{y \in \mathbb{R}} \left(\sum_{k=1}^{\infty} \left(\check{K}^{(+),\varepsilon_c} \right)^{*k}_{0,dy}(n) \right) \cdot K^{(+),\varepsilon_c}_{y,\{0\}}(1)$$

$$= 1 + \sum_{n=1}^{\infty} n \int_{y \in \mathbb{R}} \left(\sum_{k=1}^{\infty} \sum_{\substack{r_1,\ldots,r_k \in \mathbb{N}_0 \\ r_1+\cdots+r_k=n}} \left(\check{K}^{(+),\varepsilon_c}(r_1) \circ \cdots \circ \check{K}^{(+),\varepsilon_c}(r_k) \right)_{0,dy} \right) \cdot K^{(+),\varepsilon_c}_{y,\{0\}}(1)$$

$$= 1 + \sum_{k=1}^{\infty} \int_{y \in \mathbb{R}} \sum_{i=1}^{k} \left(\left(\check{D}^{(+),\varepsilon_c} \right)^{\circ(i-1)} \circ \check{G}^{(+),\varepsilon_c} \circ \left(\check{D}^{(+),\varepsilon_c} \right)^{\circ(k-i)} \right)_{0,dy} \cdot K^{(+),\varepsilon_c}_{y,\{0\}}(1)$$

$$= 1 + \int_{y \in \mathbb{R}} \left(\left(1 - \check{D}^{(+),\varepsilon_c} \right)^{-1} \circ \check{G}^{(+),\varepsilon_c} \circ \left(1 - \check{D}^{(+),\varepsilon_c} \right)^{-1} \right)_{0,dy} \cdot K^{(+),\varepsilon_c}_{y,\{0\}}(1)$$

$$= 1 + \left(\left(1 - \check{D}^{(+),\varepsilon_c} \right)^{-1} \circ \check{G}^{(+),\varepsilon_c} \circ \left(1 - \check{D}^{(+),\varepsilon_c} \right)^{-1} \circ D^{(+),\varepsilon_c} \right)_{0,\{0\}}, \tag{5.6}$$

because $K^{(+),\varepsilon_c}_{y,\{0\}}(1) = D^{(+),\varepsilon_c}_{y,\{0\}}$ and the last but one equation is due to

$$\sum_{k=1}^{\infty} \sum_{i=1}^{k} \left(\left(\check{D}^{(+),\varepsilon_c} \right)^{\circ(i-1)} \circ \check{G}^{(+),\varepsilon_c} \circ \left(\check{D}^{(+),\varepsilon_c} \right)^{\circ(k-i)} \right)_{0,dy}$$

$$= \sum_{k=1}^{\infty} \left(\left(\check{D}^{(+),\varepsilon_c} \right)^{\circ 0} \circ \check{G}^{(+),\varepsilon_c} \circ \left(\check{D}^{(+),\varepsilon_c} \right)^{\circ(k-1)} \right)_{0,dy}$$

$$+ \sum_{k=2}^{\infty} \left(\left(\check{D}^{(+),\varepsilon_c} \right)^{\circ 1} \circ \check{G}^{(+),\varepsilon_c} \circ \left(\check{D}^{(+),\varepsilon_c} \right)^{\circ(k-2)} \right)_{0,dy}$$

$$+ \sum_{k=3}^{\infty} \left(\left(\check{D}^{(+),\varepsilon_c} \right)^{\circ 2} \circ \check{G}^{(+),\varepsilon_c} \circ \left(\check{D}^{(+),\varepsilon_c} \right)^{\circ(k-3)} \right)_{0,dy} + \cdots$$

$$= \left(\sum_{i=0}^{\infty} \left(\check{D}^{(+),\varepsilon_c} \right)^{\circ i} \circ \check{G}^{(+),\varepsilon_c} \circ \sum_{k=0}^{\infty} \left(\check{D}^{(+),\varepsilon_c} \right)^{\circ k} \right)_{0,dy}.$$

Similarly to Remark 1.18 and Remark 3.23 we can see that the Markov chain $\{J_k\}_{k \geq 0}$ with the transition kernel $D^{(+),\varepsilon_c}_{x,dy}$ under $\mathcal{P}_{\varepsilon_c}^{(+)}$ is positive recurrent. Its invariant probability measure $\kappa_{\varepsilon_c}^{(+)}$ is given by $\kappa_{\varepsilon_c}^{(+)}(dx) = \nu_{\varepsilon_c}^{(+)}(x) w_{\varepsilon_c}^{(+)}(x) \mu(dx)$. The point 0 is an atom for the Markov chain, because $\kappa_{\varepsilon_c}^{(+)}(\{0\}) > 0$. Following [10], based on [25], for all $x,y \in \mathbb{R}^d$ this yields

$$\left(1 - \check{D}^{(+),\varepsilon_c} \right)^{-1}_{0,dx} = \frac{\kappa_{\varepsilon_c}^{(+)}(dx)}{\kappa_{\varepsilon_c}^{(+)}(\{0\})} \quad \text{and} \quad \left(\left(1 - \check{D}^{(+),\varepsilon_c} \right)^{-1} \circ D^{(+),\varepsilon_c} \right)_{y,\{0\}} = 1 \ .$$

Therefore by (5.6) we obtain for the first moment of η_1

$$\mathbb{E}_{\mathcal{P}_{\varepsilon_c}^{(+)}}[\eta_1] = 1 + \int_{y \in \mathbb{R}} \left(\left(1 - \breve{D}^{(+),\varepsilon_c}\right)^{-1} \circ \breve{G}^{(+),\varepsilon_c}\right)_{0,dy} \left(\left(1 - \breve{D}^{(+),\varepsilon_c}\right)^{-1} \circ D^{(+),\varepsilon_c}\right)_{y,\{0\}}$$

$$= 1 + \int_{y \in \mathbb{R}} \int_{x \in \mathbb{R}} \left(1 - \breve{D}^{(+),\varepsilon_c}\right)^{-1}_{0,dx} \breve{G}^{(+),\varepsilon_c}_{x,dy}$$

$$= 1 + \int_{y \in \mathbb{R}} \int_{x \in \mathbb{R}} \frac{\kappa_{\varepsilon_c}^{(+)}(dx)}{\kappa_{\varepsilon_c}^{(+)}(\{0\})} \varepsilon_c^{(+)} \left(\sum_{n \in \mathbb{N}} n \widetilde{f}_{x,y}^{(+)}(n)\right) \frac{\nu_{\varepsilon_c}^{(+)}(y)}{\nu_{\varepsilon_c}^{(+)}(x)} \mu(dy) .$$

Now in the pinning case it is $\widetilde{f}_{x,y}(n) \leq \text{const.} \, n^{-d/2}$ and so by lemma 5.4 one sees the finiteness of the first moment by estimating from above for $d \geq 5$

$$\mathbb{E}_{\mathcal{P}_{\varepsilon_c}}[\eta_1] \leq 1 + \text{const.} \frac{\varepsilon_c}{\nu_{\varepsilon_c}(0)w_{\varepsilon_c}(0)} \sum_{n \in \mathbb{N}} n^{\frac{2-d}{2}} \int_{y \in \mathbb{R}} \int_{x \in \mathbb{R}} \frac{\kappa_{\varepsilon_c}(dx)\nu_{\varepsilon_c}(y)}{\nu_{\varepsilon_c}(x)} \mu(dy)$$

$$\leq 1 + \text{const.} \frac{\varepsilon_c}{\nu_{\varepsilon_c}(0)w_{\varepsilon_c}(0)} \sum_{n \in \mathbb{N}} n^{\frac{2-d}{2}} \|\nu_{\varepsilon_c}\|_{L^1(\mathbb{R},d\mu)} \|w_{\varepsilon_c}\|_{L^1(\mathbb{R},d\mu)} < \infty .$$

Whereas in the wetting case by (4.9), Lemma 3.21 and Proposition 3.15 for $d \geq 3$

$$\mathbb{E}_{\mathcal{P}_{\varepsilon_c}^+}[\eta_1] \leq 1 + \text{const.} \frac{\varepsilon_c^+}{\nu_{\varepsilon_c}^+(0)w_{\varepsilon_c}^+(0)} \sum_{n \in \mathbb{N}} \frac{\log n}{n^{d/2}} \int_{y \in \mathbb{R}} \int_{x \in \mathbb{R}} (1 + cy)^2 \frac{\kappa_{\varepsilon_c}^+(dx)\nu_{\varepsilon_c}^+}{\nu_{\varepsilon_c}^+(x)} \mu(dy)$$

$$\leq 1 + \text{const.} \frac{\varepsilon_c^+}{\nu_{\varepsilon_c}^+(0)w_{\varepsilon_c}^+(0)} \sum_{n \in \mathbb{N}} \frac{\log n}{n^{d/2}} \|(1 + c\cdot)^2 \nu_{\varepsilon_c}^+\|_{L^1(\mathbb{R},d\mu)} \|w_{\varepsilon_c}^+\|_{L^1(\mathbb{R},d\mu)} < \infty .$$

The finiteness of $\|(1 + c\cdot)^2 \nu_{\varepsilon_c}^+\|_{L^1(\mathbb{R},d\mu)}$ can be seen as follows. Apply the proof of Lemma 5.4 to (5.2) with the additional factor $(1 + cx)^2$ and observe that by Lemma 3.21

$$\left\|\int_{\mathbb{R}^d} (1 + c\cdot)^2 \widetilde{f}_{x,\cdot}^+(n) \mu(dx)\right\|^2_{L^2(\mathbb{R}^d, d\mu)} \leq \text{const.} \frac{\log n}{n^{(d+4)/4}} .$$

The next step in the proof is to apply the Renewal theorem with finite mean to show

$$\liminf_{N \to \infty} \mathbb{E}_{\mathbb{P}_{\varepsilon_c,N,d}^{(+)}}\left[\frac{\ell_N}{N}\right] > 0 . \tag{5.7}$$

Setting $\iota_N := \max\{i \geq 0 \mid \eta_1 \leq N\}$ and using the higher dimensional analogues of Proposition 1.15 and Proposition 3.20, by monotoncity it suffices to show

$$\liminf_{N \to \infty} \mathbb{E}_{\mathbb{P}_{\varepsilon_c,N,d}^{(+)}}\left[\frac{\iota_N}{N}\right] = \liminf_{N \to \infty} \mathbb{E}_{\mathcal{P}_{\varepsilon_c}^{(+)}}\left[\frac{\iota_N}{N} \mid \mathcal{A}_N\right] > 0 , \tag{5.8}$$

since $\iota_N \leq \ell_N$. We have seen that $\{\eta_k\}_{k \geq 0}$ is an aperiodic renewal process, therefore by the strong LLN and the considerations on the first moment

$$\frac{\iota_N}{N} \xrightarrow[N \to \infty]{} \left(\mathbb{E}_{\mathcal{P}_{\varepsilon_c}^{(+)}}[\eta_1]\right)^{-1} , \quad \mathcal{P}_{\varepsilon_c}^{(+)}\text{-a.s.}$$

5.2 First order phase transition

and by the Renewal Theorem

$$\mathcal{P}^{(+)}_{\varepsilon_c}(\mathcal{A}_N) \xrightarrow[N\to\infty]{} \left(\mathbb{E}_{\mathcal{P}^{(+)}_{\varepsilon_c}}[\eta_1]\right)^{-1} > 0 \ .$$

Therefore

$$\mathbb{E}_{\mathcal{P}^{(+)}_{\varepsilon_c}}\left[\frac{\iota_N}{N} \mid \mathcal{A}_N\right] \xrightarrow[N\to\infty]{} \left(\mathbb{E}_{\mathcal{P}^{(+)}_{\varepsilon_c}}[\eta_1]\right)^{-1} > 0$$

and so we see that this yields (5.8). Furthermore, setting $F_N^{(+)}(\varepsilon) := \frac{1}{N}\log\frac{Z_{\varepsilon,N,d}^{(+)}}{Z_{0,N,d}^{(+)}}$, by Lemma 5.3 we have for every $\varepsilon > 0$

$$\liminf_{N\to\infty}\left(F_N^{(+)}\right)'(\varepsilon) = \frac{1}{\varepsilon}\liminf_{N\to\infty}\mathbb{E}_{\mathrm{P}^{(+)}_{\varepsilon,N,d}}\left[\frac{\ell_N}{N}\right] > 0 \ . \tag{5.9}$$

At a first glance, one could be now inclined to conclude the statement of Theorem 5.1, since of course $F_N^{(+)}(\varepsilon) \to F^{(+)}(\varepsilon)$ for all $\varepsilon > 0$. However we should be careful, since we expect the free energy not to be differentiable at criticality and so (5.9) can't be directly applied for $\varepsilon = \varepsilon_c^{(+)}$ to finish the proof. Nevertheless the properties in chapter 0 of $\widetilde{F}_N^{(+)}(t) = F_N^{(+)}(e^t)$, $t \in \mathbb{R}$ are enough to finish. Namely, since $\widetilde{F}_N^{(+)}$ is convex and non-decreasing in t we have for $h > 0$

$$\frac{\widetilde{F}_N^{(+)}(t_c+h) - \widetilde{F}_N^{(+)}(t_c)}{h} \geq \lim_{h\searrow 0}\frac{\widetilde{F}_N^{(+)}(t_c+h) - \widetilde{F}_N^{(+)}(t_c)}{h} \ , \tag{5.10}$$

where t_c is chosen such that $e^{t_c} = \varepsilon_c$. Now for the r.h.s. of (5.10) we have by (5.9)

$$\liminf_{N\to\infty}\lim_{h\searrow 0}\frac{\widetilde{F}_N^{(+)}(t_c+h) - \widetilde{F}_N^{(+)}(t_c)}{h} = \liminf_{N\to\infty}\lim_{h\searrow 0}\left(\frac{(e^h-1)e^{t_c}}{h}\frac{F_N^{(+)}(e^{t_c+h}) - F_N^{(+)}(e^{t_c})}{e^{t_c+h} - e^{t_c}}\right)$$
$$= \varepsilon_c\liminf_{N\to\infty}\left(F_N^{(+)}\right)'(\varepsilon_c) > 0 \ .$$

The inequality (5.10) is true for every $h > 0$, so we can first apply the limit $N \to \infty$ and then on the l.h.s. $h \searrow 0$, i.e.

$$\lim_{h\searrow 0}\liminf_{N\to\infty}\frac{\widetilde{F}_N^{(+)}(t_c+h) - \widetilde{F}_N^{(+)}(t_c)}{h} = \lim_{h\searrow 0}\frac{\widetilde{F}^{(+)}(t_c+h) - \widetilde{F}^{(+)}(t_c)}{h}$$
$$= \lim_{h\searrow 0}\left(\frac{(e^h-1)e^{t_c}}{h}\frac{F^{(+)}(e^{t_c+h}) - F^{(+)}(e^{t_c})}{e^{t_c+h} - e^{t_c}}\right) \ .$$

Altogether this yields

$$\varepsilon_c\lim_{h\searrow 0}\frac{F^{(+)}(\varepsilon_c+h)}{h} \geq \varepsilon_c\liminf_{N\to\infty}\left(F_N^{(+)}\right)'(\varepsilon_c) > 0 \ ,$$

which is exactly the statement of the Theorem 5.1.

□

Remark 5.6
At this point we remark that in the Gaussian Laplacian pinning-model already for $d \geq 2$ a first order phase transition occurs. We don't want to go into detail, but considering Remark 4.4.4 and the last proof, on can show that

$$||\nu_\varepsilon||_{L^2(\mathbb{R}^d,d\mu)} \leq \text{const.} ||\nu_\varepsilon||_{L^2(\mathbb{R}^d,d\mu)} \sum_{n=1}^{\infty} \frac{1}{n^{3d/2}} < \infty$$

and also the same for the left eigenvalue w_ε. Furthermore it holds for $d \geq 2$

$$\mathbb{E}_{\mathcal{P}_{\varepsilon_c}}[\eta_1] \leq 1 + \text{const.} \frac{\varepsilon_c}{\nu_{\varepsilon_c}(0) w_{\varepsilon_c}(0)} \sum_{n \in \mathbb{N}} n^{1-2d} ||\nu_{\varepsilon_c}||_{L^2(\mathbb{R},d\mu)} ||w_{\varepsilon_c}||_{L^2(\mathbb{R},d\mu)} < \infty .$$

5.3 Smooth phase transition

Now we are going to investigate the lower dimensions, i.e. we will prove Theorem 5.2. Here one has to distinguish between two different cases, the case of a "proper" and the case of a trivial phase transition.

Observe that by the proof of Lemma 3.21 we have for every bounded Borel set $A \in \mathcal{B}(\mathbb{R}^d)$

$$\sup_{x \in A} \widetilde{f}_{x,y}(n) \sim \frac{c}{n^{d/2}} e^{-a||y||_2^2} \quad , \text{ for every } y \in \mathbb{R}^d , \tag{5.11}$$

where $a := \lim_{n \to \infty} a(\gamma, n) = \frac{1-\gamma^2}{\sigma^2} > 0$ and for wetting we assume the condition

$$(\mathbf{CW}) \qquad \sup_{x \in A} \widetilde{f}^+_{x,y}(n) \succeq \frac{c^+(y)}{n^{(d+2)/2}} \quad , \text{ for every } y \in \mathbb{R}^d ,$$

where $c^+(y)$ is exponentially decreasing.

Remark 5.7
Let us make a short comment on the assumption $(\mathbf{CW})$. First of all it will be needed in order to use Karamata's Tauberian Theorem in (5.22). Especially for the wetting model in $d = 2$ there is no room for relaxing assumption $(\mathbf{CW})$, since here we will need $M_\varepsilon = \mathbb{E}_{\mathcal{P}_\varepsilon^{(+)}}(\eta_1) \to \infty$, as $\varepsilon \searrow \varepsilon_c^{(+)}$. This means that a lower bound on the conditional entropic repulsion $w_{x,y}(n)$ with exact order in n, i.e. n^{-1} is needed. In view of Proposition 3.6 one should believe that a proof for the lower bound in entropic repulsion cannot be to difficult. However, like for the upper bound of $w_{x,y}(n)$ one has to deal with the conditioning and a decoupling-argument seems not to work in this case.

5.3.1 The case of "proper" phase transition

We will start by considering the case of "proper" phase transition, meaning $d = 3, 4$ for pinning and $d = 1, 2$ for the wetting model. In order to make a statement on the smooth

5.3 Smooth phase transition

phase transition we will investigate the distribution of the first double-contact η_1. Similarly to the proof of the first order phase transition we can write for $n \in \mathbb{N}$

$$\mathcal{P}_\varepsilon^{(+)}(\eta_1 = n) = \int_{y \in \mathbb{R}^d} \left(\sum_{k=0}^{\infty} \left(\breve{K}^{(+),\varepsilon} \right)_{0,dy}^{*k} (n-1) \right) \cdot K_{y,\{0\}}^{(+),\varepsilon}(1)$$

$$= \varepsilon\, e^{-F^{(+)}(\varepsilon)n} \int_{y \in \mathbb{R}^d} \left(\sum_{k=0}^{\infty} \varepsilon^k \left(\breve{\tilde{F}}^{(+)} \right)_{0,dy}^{*k} (n-1) \right) \cdot \tilde{F}_{y,\{0\}}^{(+)}(1) \,. \quad (5.12)$$

We fix now an $\varepsilon > 0$ and set

$$\mathcal{F}_{x,dy}^{(+)}(n) := \varepsilon \breve{\tilde{F}}_{x,dy}^{(+)}(n) \,.$$

This definition is just because of notational reasons in further calculations. In the following we consider a lower and upper bound for (5.12), when $n \to \infty$.

Lower asymptotical bound

As a first step we show a result on the lower bound asymptotics of the distribution of η_1, i.e. Propositions 5.9 below. We are going to bound asymptotically the bracket part in (5.12) from below. For this purpose we prove first the

Lemma 5.8 *For the pining (wetting) model we have in dimensions $d \geq 3$ ($d \geq 1$)*

$$\left(\mathcal{F}^{(+)} \right)_{x,dy}^{*k} (n) \succeq \frac{1}{g^{(+)}(n)} \sum_{i=0}^{k-1} \left(\left(\mathcal{G}^{(+)} \right)^{\circ i} \circ \mathcal{L} \circ \left(\mathcal{G}^{(+)} \right)^{\circ (k-1-i)} \right)_{x,dy}$$

for all $x, y \in \mathbb{R}^d$, where

$$g^{(+)}(n) = \begin{cases} n^{d/2} & \text{, for pinning} \\ n^{(d+2)/2} & \text{, for wetting} \end{cases}$$

and

$$\mathcal{G}_{x,dy}^{(+)} := \sum_{n \in \mathbb{N}} \mathcal{F}_{x,dy}^{(+)} \,, \quad \mathcal{L}_{x,dy} := \varepsilon \min\{c\, e^{-ay^2}, c^+(y)\}\, \mu(dy) \,.$$

The statement on wetting is of course only valid under assumption (**CW**).

Proof We will prove this by induction. For $k = 1$ we have on the one hand by definition, (5.11) and (**CW**)

$$\left(\mathcal{F}^{(+)} \right)_{x,dy}^{*1} (n) \succeq \frac{\mathcal{L}_{x,dy}}{g^{(+)}(n)}$$

and on the other hand

$$\left(\left(\mathcal{G}^{(+)} \right)^{\circ 0} \circ \mathcal{L} \circ \left(\mathcal{G}^{(+)} \right)^{\circ 0} \right)_{x,dy} = \int_{z \in \mathbb{R}^d} \int_{u \in \mathbb{R}^d} \delta_x(du)\, \mathcal{L}_{u,dz}\, \delta_z(dy) = \mathcal{L}_{x,dy} \,.$$

Now the step to $k+1$ can be made as follows

$$\left(\mathcal{F}^{(+)}\right)^{*(k+1)}_{x,dy}(n)$$
$$= \sum_{m=1}^{n/2} \int_{z \in \mathbb{R}^d} \left[\left(\mathcal{F}^{(+)}\right)^{*k}_{x,dz}(m) \left(\mathcal{F}^{(+)}\right)_{z,dy}(n-m) + \left(\mathcal{F}^{(+)}\right)^{*k}_{x,dz}(n-m) \left(\mathcal{F}^{(+)}\right)_{z,dy}(m)\right]$$
$$=: \sum_{m=1}^{n/2} \int_{z \in \mathbb{R}^d} [I(x,z,y,k,n,m) + II(x,z,y,k,n,m)]$$

Now by Fatou's Lemma and the induction step

$$g^{(+)}(n) \sum_{m=1}^{n/2} \int_{z \in \mathbb{R}^d} I(x,z,y,k,n,m) = \int_{z \in \mathbb{R}^d} \sum_{m=1}^{n/2} \left(\mathcal{F}^{(+)}\right)^{*k}_{x,dz}(m) \left[g^{(+)}(n) \left(\mathcal{F}^{(+)}\right)_{z,dy}(n-m)\right]$$
$$\succeq \int_{z \in \mathbb{R}^d} \liminf_{n \to \infty} \sum_{m=1}^{n/2} \left(\mathcal{F}^{(+)}\right)^{*k}_{x,dz}(m) \left(\left(\mathcal{G}^{(+)}\right)^{\circ 0} \circ \mathcal{L} \circ \left(\mathcal{G}^{(+)}\right)^{\circ 0}\right)_{z,dy}$$
$$= \int_{z \in \mathbb{R}^d} \left(\mathcal{G}^{(+)}\right)^{\circ k}_{x,dz} \mathcal{L}_{z,dy} = \left[\left(\mathcal{G}^{(+)}\right)^{\circ k} \circ \mathcal{L}\right]_{x,dy}, \qquad (5.13)$$

where we have used

$$g^{(+)}(n)/g^{(+)}(n-m) \underset{n \to \infty}{\succeq} 1 \text{, for } m = 1, ..., n/2 \quad \text{and} \quad \left(\mathcal{G}^{(+)}\right)^{\circ k}_{x,dy} = \sum_{n \in \mathbb{N}} \left(\mathcal{F}^{(+)}\right)^{*k}_{x,dy}(n),$$

which can be shown by induction. Similarly we can treat the second part

$$g^{(+)}(n) \sum_{m=1}^{n/2} \int_{z \in \mathbb{R}^d} II(x,z,y,k,n,m) = \int_{z \in \mathbb{R}^d} \sum_{m=1}^{n/2} \left[g^{(+)}(n) \left(\mathcal{F}^{(+)}\right)^{*k}_{x,dz}(n-m)\right] \left(\mathcal{F}^{(+)}\right)_{z,dy}(m)$$
$$\succeq \int_{z \in \mathbb{R}^d} \sum_{i=0}^{k-1} \left(\left(\mathcal{G}^{(+)}\right)^{\circ i} \circ \mathcal{L} \circ \left(\mathcal{G}^{(+)}\right)^{\circ(k-1-i)}\right)_{x,dz} \lim_{n \to \infty} \sum_{m=1}^{n/2} \left(\mathcal{F}^{(+)}\right)^{*k}_{z,dy}(m)$$
$$= \int_{z \in \mathbb{R}^d} \sum_{i=0}^{k-1} \left(\left(\mathcal{G}^{(+)}\right)^{\circ i} \circ \mathcal{L} \circ \left(\mathcal{G}^{(+)}\right)^{\circ(k-1-i)}\right)_{x,dz} \left(\mathcal{G}^{(+)}\right)^{*k}_{z,dy}$$
$$= \sum_{i=0}^{k-1} \left(\left(\mathcal{G}^{(+)}\right)^{\circ i} \circ \mathcal{L} \circ \left(\mathcal{G}^{(+)}\right)^{\circ(k-i)}\right)_{x,dy}, \qquad (5.14)$$

Putting (5.13) and (5.14) together we obtain

$$g^{(+)}(n) \left(\mathcal{F}^{(+)}\right)^{*(k+1)}_{x,dy}(n) \succeq \sum_{i=0}^{k} \left(\left(\mathcal{G}^{(+)}\right)^{\circ i} \circ \mathcal{L} \circ \left(\mathcal{G}^{(+)}\right)^{\circ(k-i)}\right)_{x,dy}.$$

$\square$

5.3 Smooth phase transition

We can now use Fatou's Lemma to obtain by Lemma 5.8

$$g^{(+)}(n) \sum_{k=0}^{\infty} \left(\mathcal{F}^{(+)}\right)^{*k}_{x,dy}(n) \succeq \sum_{k=0}^{\infty} \sum_{i=0}^{k-1} \left(\left(\mathcal{G}^{(+)}\right)^{\circ i} \circ \mathcal{L} \circ \left(\mathcal{G}^{(+)}\right)^{\circ(k-1-i)}\right)_{x,dy}$$

$$= \left[\left(1 - \mathcal{G}^{(+)}\right)^{-1} \circ \mathcal{L} \circ \left(1 - \mathcal{G}^{(+)}\right)^{-1}\right]_{x,dy}$$

$$= \varepsilon \left[\left(1 - \varepsilon \check{\mathcal{G}}^{(+)}\right)^{-1} \circ \check{\mathcal{L}} \circ \left(1 - \varepsilon \check{\mathcal{G}}^{(+)}\right)^{-1}\right]_{x,dy}$$

where

$$\check{\mathcal{G}}^{(+)}_{x,dy} := \sum_{n \in \mathbb{N}} \check{F}^{(+)}_{x,dy}(n) \quad \text{and} \quad \check{\mathcal{L}}^{(+)}_{x,dy} := \min\left\{c\, e^{-ay^2}, c^+(y)\right\} \mu(dy) .$$

Now setting

$$c_1(\varepsilon) := \varepsilon^2 \int_{y \in \mathbb{R}^d} \left[\left(1 - \varepsilon \check{\mathcal{G}}^{(+)}\right)^{-1} \circ \check{\mathcal{L}} \circ \left(1 - \varepsilon \check{\mathcal{G}}^{(+)}\right)^{-1}\right]_{0,dy} \cdot \widetilde{F}^{(+)}_{y,\{0\}}(1) .$$

By (5.12) and Fatou's Lemma we finally obtain the

Proposition 5.9 *For every $\varepsilon > 0$ (and under assumption (CW) for wetting) it holds*

$$\mathcal{P}^{(+)}_\varepsilon(\eta_1 = n) \succeq \frac{c_1(\varepsilon)}{g^{(+)}(n)} e^{-F^{(+)}(\varepsilon)n} \quad , \text{ for } c_1(\varepsilon) > 0 .$$

Upper asymptotical bound

The upper bound for (5.12), i.e. Proposition 5.11, can be obtained in a similar manner to the lower bound. However we have to be careful, since we can't use Fatou's Lemma. Therefore one has to work a little bit and use the dominated convergence Theorem. First of all we have in the pinning case (5.11) and by (1.21) there is an $\tilde{c} > 0$, s.th. for all $x, y \in \mathbb{R}^d$ and $n \in \mathbb{N}$

$$\widetilde{f}_{x,y}(n) \leq \frac{\tilde{c}}{n^{d/2}} . \tag{5.15}$$

Whereas in the wetting case, due to Proposition 3.15 and Lemma 3.21, we can find an $c^+ > 0$, s.th. for all $x, y \subset \mathbb{R}^d$ and $n \subset \mathbb{N}$

$$\widetilde{f}^+_{x,y}(n) \leq \frac{c^+ \log n}{n^{(d+2)/2}} . \tag{5.16}$$

Let us start by bounding asymptotically the bracket part in (5.12) from above.

Lemma 5.10 *For the pining (wetting) model we have in dimensions $d \geq 3$ ($d \geq 1$) for all $k, n \in \mathbb{N}$ and $x, y \in \mathbb{R}^d$*

$$\left(\mathcal{F}^{(+)}\right)^{*k}_{x,dy}(n) \leq \frac{g(k)}{\tilde{g}^{(+)}(n)} \sum_{i=0}^{k-1} \left(\left(\mathcal{G}^{(+)}\right)^{\circ i} \circ \hat{\mathcal{L}} \circ \left(\mathcal{G}^{(+)}\right)^{\circ(k-1-i)}\right)_{x,dy} ,$$

where $\tilde{g}^+(n) := g^+(n)/\log n$ and $\tilde{g} \equiv g$ and $\hat{\mathcal{L}}_{x,dy} := \varepsilon \max\{\tilde{c}, c^+\} \mu(dy) .$

Proof We will prove this again by induction. For $k = 1$ we have on the one hand by (5.15) and (5.16)
$$\left(\mathcal{F}^{(+)}\right)_{x,dy}^{*1}(n) \le \frac{\hat{\mathcal{L}}_{x,dy}}{\tilde{g}^{(+)}(n)}$$
and on the other hand $g(1) = 1$ and
$$\left(\left(\mathcal{G}^{(+)}\right)^{\circ 0} \circ \hat{\mathcal{L}} \circ \left(\mathcal{G}^{(+)}\right)^{\circ 0}\right)_{x,dy} = \int_{z \in \mathbb{R}^d} \int_{u \in \mathbb{R}^d} \delta_x(du)\, \hat{\mathcal{L}}_{u,dz}\, \delta_z(dy) = \hat{\mathcal{L}}_{x,dy}\,.$$
Wlog we treat the induction step in the even case (the odd case is analogous)
$$\left(\mathcal{F}^{(+)}\right)_{x,dy}^{*(2k)}(n)$$
$$\le \int_{z \in \mathbb{R}^d} \sum_{m=1}^{\lceil n/2 \rceil} \left[\left(\mathcal{F}^{(+)}\right)_{x,dz}^{*k}(m) \left(\mathcal{F}^{(+)}\right)_{z,dy}^{*k}(n-m) + \left(\mathcal{F}^{(+)}\right)_{x,dz}^{*k}(n-m) \left(\mathcal{F}^{(+)}\right)_{z,dy}^{*k}(m) \right]$$
$$\le \frac{g(k)}{\tilde{g}^{(+)}(n/2)} \left[\int_{z \in \mathbb{R}^d} \sum_{m=1}^{\infty} \left(\mathcal{F}^{(+)}\right)_{x,dz}^{*k}(m) \sum_{i=0}^{k-1} \left(\left(\mathcal{G}^{(+)}\right)^{\circ i} \circ \hat{\mathcal{L}} \circ \left(\mathcal{G}^{(+)}\right)^{\circ(k-1-i)}\right)_{z,dy} \right.$$
$$\left. + \int_{z \in \mathbb{R}^d} \sum_{i=0}^{k-1} \left(\left(\mathcal{G}^{(+)}\right)^{\circ i} \circ \hat{\mathcal{L}} \circ \left(\mathcal{G}^{(+)}\right)^{\circ(k-1-i)}\right)_{x,dz} \sum_{m=1}^{\infty} \left(\mathcal{F}^{(+)}\right)_{z,dy}^{*k}(m) \right]$$
$$= \frac{g(k)}{\tilde{g}^{(+)}(n/2)} \left[\sum_{i=0}^{k-1} \left(\left(\mathcal{G}^{(+)}\right)^{\circ(k+i)} \circ \hat{\mathcal{L}} \circ \left(\mathcal{G}^{(+)}\right)^{\circ(k-1-i)}\right)_{x,dy} \right.$$
$$\left. + \sum_{i=0}^{k-1} \left(\left(\mathcal{G}^{(+)}\right)^{\circ i} \circ \hat{\mathcal{L}} \circ \left(\mathcal{G}^{(+)}\right)^{\circ(2k-1-i)}\right)_{x,dy} \right]$$
$$\le \frac{g(2k)}{\tilde{g}^{(+)}(n)} \sum_{i=0}^{2k-1} \left(\left(\mathcal{G}^{(+)}\right)^{\circ i} \circ \hat{\mathcal{L}} \circ \left(\mathcal{G}^{(+)}\right)^{\circ(2k-1-i)}\right)_{x,dy},$$
since
$$\frac{g(k)}{\tilde{g}^+(n/2)} = \frac{k^{(d+2)/2}}{(n/2)^{(d+2)/2}/\log(n/2)} = \frac{(2k)^{(d+2)/2} \log(n/2)}{n^{(d+2)/2}} \le \frac{g(2k)}{\tilde{g}^+(n)}$$
and analogously for the pinning case. □

Next similarly to [10] we can find an $a > 0$, such that for every $\varepsilon \in [\varepsilon_c^{(+)}, \varepsilon_c^{(+)} + a]$ the spectral radius $\varrho(\varepsilon)$ of $\varepsilon \check{\mathcal{G}}_{x,dy}^{(+)}$ (recall $\check{\mathcal{G}}_{x,dy}^{(+)} = B_{x,dy}^{(+),0} \mathbb{1}_{y \ne 0}$) is strictly smaller than one. At this point we mention that Lemma 4.2 in [10] can be also used to operators on $L^p(\mathbb{R}, d\mu)$ for $1 \le p \le \infty$, this is needed for the pinning case in $d = 1, 2$, cf. Proposition 4.8. We will show that an $\beta > 1$ can be chosen in such a way that for all $\varepsilon \in [\varepsilon_c^{(+)}, \varepsilon_c^{(+)} + a]$ we have
$$\int_{y \in \mathbb{R}^d} \left(1 - \beta \varepsilon \check{\mathcal{G}}^{(+)}\right)_{x,dy}^{-1} \left(\int_{z \in \mathbb{R}^d} \mu(dz) \int_{u \in A} \left(1 - \beta \varepsilon \check{\mathcal{G}}^{(+)}\right)_{z,du}^{-1}\right) < \infty\,. \quad (5.17)$$
For this purpose we choose an $\beta > 1$ such that $\sup\{\varrho(\varepsilon\beta) \mid \varepsilon \in [\varepsilon_c^{(+)}, \varepsilon_c^{(+)} + a]\} < 1$. By [21] III-6.2 we obtain
$$\left\| \left(\varepsilon\beta\check{\mathcal{G}}^{(+)}\right)^{\circ k} \right\|_{L^p(\mathbb{R},d\mu)}^{1/k} \xrightarrow{k \to \infty} \varrho(\varepsilon\beta) < 1 \quad (5.18)$$

5.3 Smooth phase transition

where $p=1$ for pinning and $p=2$ for the wetting case. We consider first the $\mu(dy)$-integral in (5.17) and by splitting the first summand and applying Hölder's inequality we have

$$\int_{y\in\mathbb{R}^d}\left(1-\beta\varepsilon\breve{\tilde{\mathcal{G}}}^{(+)}\right)^{-1}_{x,dy}\leq 1+\varepsilon\beta\sum_{k=0}^{\infty}\left\|\left(\varepsilon\beta\breve{\tilde{\mathcal{G}}}^{(+)}\right)^{\circ k}\right\|_{L^p(\mathbb{R},d\mu)}\left\|\int_{y\in\mathbb{R}^d}\breve{\tilde{\mathcal{G}}}^{(+)}_{\cdot,dy}\right\|_{L^q(\mathbb{R},d\mu)}, \quad (5.19)$$

where $q=\infty$ for pinning and $q=2$ for the wetting case. Apparently by (5.18) the equation (5.19) is finite, if $\|\cdots\|_{L^q(\mathbb{R},d\mu)}$ is bounded. Indeed, in the pinning case by (4.12)

$$\left\|\int_{y\in\mathbb{R}^d}\breve{\tilde{\mathcal{G}}}^{(+)}_{\cdot,dy}\right\|_{L^{\infty}(\mathbb{R},d\mu)}\leq\sum_{n\in\mathbb{N}}\int_{y\in\mathbb{R}^d}\left\|\widetilde{F}_{z,dy}(n)\mathbb{1}_{y\neq 0}\right\|_{L^{\infty}(\mathbb{R},d\mu)}\leq\sum_{n=2}^{\infty}\frac{\text{const.}}{n^{d/2}}<\infty.$$

The wetting case can be checked in a similar way to the proof of Proposition 3.22. Therefore (5.19) is finite and so is (5.17), since the second integral in (5.17) can be treated in a similar way to the first one. We take a constant c such that $g(k)\leq c\beta^k$ for all $k\in\mathbb{N}$. Then by (5.17) for every $x\in\mathbb{R}^d$ and bounded $A\in\mathcal{B}(\mathbb{R}^d)$

$$\sum_{k=1}^{\infty}\sum_{i=0}^{k-1}g(k)\left(\left(\mathcal{G}^{(+)}\right)^{\circ i}\circ\hat{\mathcal{L}}\circ\left(\mathcal{G}^{(+)}\right)^{\circ(k-1-i)}\right)_{x,A}$$
$$\leq c\beta\max\{\tilde{c},c^+\}\varepsilon\int_{y\in\mathbb{R}^d}\left(1-\beta\varepsilon\breve{\tilde{\mathcal{G}}}^{(+)}\right)^{-1}_{x,dy}\left(\int_{z\in\mathbb{R}^d}\mu(dz)\int_{u\in A}\left(1-\beta\varepsilon\breve{\tilde{\mathcal{G}}}^{(+)}\right)^{-1}_{z,du}\right)<\infty$$
$$(5.20)$$

Now following the proof of Lemma 5.8 and using Lemma 5.10 and (5.20) for the dominated convergence Theorem instead of Fatou's Lemma, we obtain for all $x,y\in\mathbb{R}^d$

$$\left(\mathcal{F}^{(+)}\right)^{*k}_{x,dy}(n)\preceq\frac{1}{\tilde{g}^{(+)}(n)}\sum_{i=0}^{k-1}\left(\left(\mathcal{G}^{(+)}\right)^{\circ i}\circ\hat{\mathcal{L}}\circ\left(\mathcal{G}^{(+)}\right)^{\circ(k-1-i)}\right)_{x,dy} \quad (5.21)$$

Again, thanks to Lemma 5.10 and (5.20) we can apply the dominated convergence Theorem and then (5.21) to get

$$\tilde{g}^{(+)}(n)\sum_{k=0}^{\infty}\left(\mathcal{F}^{(+)}\right)^{*k}_{x,dy}(n)\preceq\sum_{k=0}^{\infty}\sum_{i=0}^{k-1}\left(\left(\mathcal{G}^{(+)}\right)^{\circ i}\circ\mathcal{L}\circ\left(\mathcal{G}^{(+)}\right)^{\circ(k-1-i)}\right)_{0,dy}$$
$$=\max\{\tilde{c},c^+\}\varepsilon\int_{u\in\mathbb{R}^d}\left(1-\varepsilon\breve{\tilde{\mathcal{G}}}^{(+)}\right)^{-1}_{0,du}\left(\int_{w\in\mathbb{R}^d}\mu(dw)\left(1-\varepsilon\breve{\tilde{\mathcal{G}}}^{(+)}\right)^{-1}_{w,dy}\right)$$

Let us set

$$c_2(\varepsilon):=\varepsilon^2\max\{c,c^+\}\int_{u\in\mathbb{R}^d}\left(1-\varepsilon\breve{\tilde{\mathcal{G}}}^{(+)}\right)^{-1}_{0,du}\left(\int_{w\in\mathbb{R}^d}\mu(dw)\int_{y\in\mathbb{R}^d}\left(1-\varepsilon\breve{\tilde{\mathcal{G}}}^{(+)}\right)^{-1}_{w,dy}\cdot\widetilde{F}^{(+)}_{y,\{0\}}(1)\right).$$

Then again, due to Lemma 5.10 and (5.20) we can apply the dominated convergence Theorem the third time to conclude from (5.12)

Proposition 5.11 *There is an $a > 0$ such that for every $\varepsilon_c^{(+)} \leq \varepsilon \leq \varepsilon_c^{(+)} + a$ it holds*

$$\mathcal{P}_\varepsilon^{(+)}(\eta_1 = n) \preceq \frac{c_2(\varepsilon)}{\bar{g}^{(+)}(n)} e^{-F^{(+)}(\varepsilon)n} \quad, \text{ for } c_2(\varepsilon) > 0 \, .$$

Remark 5.12
We believe that in wetting case the asymptotical lower bound in Proposition 5.9 describes also the true asymptotical behavior. The additional term $\log n$ in the upper bound of Proposition 5.11 is due to Proposition 3.15. Observe that in the pinning case we have the same asymptotical behavior for lower and upper bound. One could even improve to obtain constants $c_1 \equiv c_2$, but this is of no further interest.

5.3.2 Proof of Theorem 5.2 in case of "proper" phase transition

Recall that we are still treating the case of $d = 3, 4$ for pinning and $d = 1, 2$ for wetting. Consider $\varepsilon \in (\varepsilon_c^{(+)}, \varepsilon_c^{(+)} + a]$. Then by Proposition 5.11 we know that

$$M_\varepsilon := \mathbb{E}_{\mathcal{P}_\varepsilon^{(+)}}(\eta_1) < \infty \, .$$

Thanks to Proposition 5.9, Karamata's Tauberian Theorem ([4] page 37) yields for an $c > 0$

$$M_\varepsilon \succeq c \begin{cases} (F^{(+)}(\varepsilon))^{-1/2} & \text{, for pinning}(d = 3) \text{ and wetting } (d = 1) \\ \log (F^{(+)}(\varepsilon)^{-1}) & \text{, for pinning } (d = 4) \text{ and wetting } (d = 2) \, , \end{cases} \quad (5.22)$$

as $\varepsilon \searrow \varepsilon_c^{(+)}$. Now following [10] we can obtain

$$\limsup_{N\to\infty} \mathbb{E}_{\mathcal{P}_\varepsilon^{(+)}} \left[\frac{\ell_N}{N} \Big| \mathcal{A}_N \right] \leq \frac{4}{\nu_\varepsilon^{(+)}(\{0\}) M_\varepsilon} \xrightarrow[\varepsilon \searrow \varepsilon_c^{(+)}]{} 0 \, ,$$

since $\nu_{\varepsilon_c^{(+)}}^{(+)}(\{0\}) > 0$ and due to (5.22) we have $M_\varepsilon \to \infty$. Now by the higher dimensional analogous of Proposition 1.15 and Proposition 3.20 this yields

$$\limsup_{\varepsilon \searrow \varepsilon_c^{(+)}} \limsup_{N\to\infty} \mathbb{E}_{\mathbb{P}_{\varepsilon,N,d}^{(+)}} \left[\frac{\ell_N}{N} \right] = \limsup_{\varepsilon \searrow \varepsilon_c^{(+)}} \limsup_{N\to\infty} \mathbb{E}_{\mathcal{P}_\varepsilon^{(+)}} \left[\frac{\ell_N}{N} \Big| \mathcal{A}_N \right] = 0 \, . \quad (5.23)$$

Recall the notation of $F_N^{(+)}(\varepsilon)$ in this chapter, then by (5.23) and Lemma 5.3 we have

$$\limsup_{\varepsilon \searrow \varepsilon_c^{(+)}} \limsup_{N\to\infty} \left(F_N^{(+)} \right)'(\varepsilon) = 0 \, . \quad (5.24)$$

Now since the function $\widetilde{F}_N^{(+)}(t) = F_N^{(+)}(e^t)$, $t \in \mathbb{R}$ is convex and non-decreasing in t, cf. the line above (5.10), we have for $e^{t_c} = \varepsilon_c^{(+)}$ and $h > 0$

$$\frac{\widetilde{F}_N^{(+)}(t_c + h) - \widetilde{F}_N^{(+)}(t_c)}{h} \leq \lim_{h_1 \searrow 0} \frac{\widetilde{F}_N^{(+)}(t_c + h + h_1) - \widetilde{F}_N^{(+)}(t_c + h)}{h_1} \, .$$

5.3 Smooth phase transition

For the l.h.s. we get

$$\lim_{h\searrow 0}\lim_{N\to\infty}\frac{\widetilde{F}_N^{(+)}(t_c+h)-\widetilde{F}_N^{(+)}(t_c)}{h}=\varepsilon_c^{(+)}\lim_{h\searrow 0}\frac{F^{(+)}(\varepsilon_c^{(+)}+h)}{h}$$

and for the r.h.s. it follows from (5.24)

$$\limsup_{h\searrow 0}\limsup_{N\to\infty}\lim_{h_1\searrow 0}\frac{\widetilde{F}_N^{(+)}(t_c+h+h_1)-\widetilde{F}_N^{(+)}(t_c+h)}{h_1}$$

$$=\limsup_{h\searrow 0}\limsup_{N\to\infty}\left(\frac{e^{t_c+h+h_1}-e^{t_c+h}}{h_1}\frac{F_N^{(+)}(e^{t_c+h+h_1})-F_N^{(+)}(e^{t_c+h})}{e^{t_c+h+h_1}-e^{t_c+h}}\right)$$

$$=\limsup_{h\searrow 0}e^{t_c+h}\limsup_{N\to\infty}\left(F_N^{(+)}\right)'(\varepsilon_c^{(+)}e^h)=0 \,,$$

which proves a smooth phase transition for $d=3,4$ in pinning and for $d=1,2$ in the wetting case.

What remains is to show that the phase transition is of second order. But since we know that under $\mathcal{P}_{\varepsilon_c^{(+)}}^{(+)}$ the process $\{\eta_i\}_{i\in\mathbb{Z}^+}$ is a classical renewal process, by Propositions 5.9 and 5.11 this can be answered in the same way as in [10]. Let us make a short comment on this. Due to [10] we have $F^{(+)}(\varepsilon_c+\varepsilon)\geq G^{(+)}(\varepsilon)$, where $G^{(+)}$ is the free energy of a classical pinning model. We know already that

$$\frac{c_-}{g^{(+)}(n)}\preceq \mathcal{P}_{\varepsilon_c}^{(+)}(\eta_1=n)\preceq \frac{c_+}{\tilde{g}^{(+)}(n)} \quad,\ n\to\infty$$

where we assumed **(CW)** for the lower bound in wetting case. Therefore by [18] Thm. 2.1 (2) it follows for pinning ($d=3$) and wetting ($d=1$)

$$c'_-\varepsilon^2 \preceq G(\varepsilon)\preceq c'_+\varepsilon^2 \ \text{and} \ \ c'_-\varepsilon^2\preceq G^+(\varepsilon)\preceq c'_+(c)\varepsilon^{2-c} \quad,\ \varepsilon\to 0$$

for any $c>0$ close to 0. For the pinning model ($d=4$) and the wetting ($d=2$) the asymptotics of $\mathcal{P}_{\varepsilon_c}^{(+)}(\eta_1=n)$ reveals the same behavior as for the Laplacian model in dimension $d=1$, cf. [10]. Therefore $G^{(+)}(\varepsilon)$ behaves up to a constant like $\varepsilon/\log(1/\varepsilon)$, $\varepsilon\to 0$. Therefore in all here considered cases similarly to [10] we obtain for $\varepsilon\to 0$

$$F^{(+)}(\varepsilon_c+\varepsilon)\succeq(\text{const.})\min\{\varepsilon^2,\varepsilon/\log(1/\varepsilon)\} \,,$$

which just means that the phase transition is of second order.

5.3.3 Theorem 5.2 and trivial phase transition

We are left with the case of pinning in dimensions $d=1$ and $d=2$, where the transition is trivial, meaning $\varepsilon_c=0$. We remark first that due to (4.23) we have a monotonicity of the free energy in the dimension, that is for all $\varepsilon\geq 0$ and $d\geq 1$ it is $F_{d+1}(\varepsilon)\leq F_d(\varepsilon)$. Consequently a smooth phase transition for $d=1$ would immediately imply a smooth phase transition for $d=2$, if $F_2(\varepsilon)$ was differentiable in 0.

Remark 5.13
Since our mixed model reveals similar behavior to the gradient model, it is worth first to take a look at it. Concerning the order of phase transition for the Gaussian gradient model we couldn't find any reference. However referring to [7] this can be done easily. Bolthausen, Funaki and Otobe showed that in this case the free energy can be expressed as $F(\epsilon) = -\log(x(\epsilon))$, where $x = x(\epsilon)$ is the solution of

$$\frac{1}{\epsilon} = \sum_{n=1}^{\infty} \frac{x^n}{(2\pi n)^{1/2}} = \sum_{n=1}^{\infty} \frac{e^{-F(\epsilon)n}}{(2\pi n)^{1/2}}$$

Now Karamata's Tauberian Theorem ([4] page 37) yields that $F(\varepsilon) \sim c\varepsilon^2$, $\varepsilon \to 0$. This means a second order phase transition. In an analogous way we have for $d = 2$

$$\frac{1}{\epsilon} = c\sum_{n=1}^{\infty} \frac{x^n}{n} = c\sum_{n=1}^{\infty} \frac{e^{-F(\epsilon)n}}{n}$$

and $F(\varepsilon) \sim \exp(-(c\varepsilon)^{-1})$, which means an infinite order of phase transition.

Let us first concentrate on the dimension one for our mixed pinning model. For this purpose let us write the product measure in (1.1) according to Lemma 1.20. Recall the partition function for the model

$$\mathcal{Z}_{\varepsilon,N} = \int_{\mathbb{R}^{N-1}} e^{-\mathcal{H}_{[-1,N]}(\varphi)} \prod_{i=1}^{N-1} (\varepsilon\delta_0(d\varphi_i) + d\varphi_i)$$

where we have $\varphi_{-1} = \varphi_0 = \varphi_N = \varphi_{N+1} = 0$ and

$$\mathcal{H}_{[-1,N+1]}(\varphi) = \alpha \sum_{i=1}^{N+1} \frac{1}{2}(\nabla\varphi_i)^2 + \beta \sum_{i=0}^{N} \frac{1}{2}(\Delta\varphi_i)^2 \ .$$

Assumption 5.14
We make now the following assumption: one can write $\delta_0(d\varphi_i)$ instead of $d\varphi_{i+1}$ in such a way that

$$\int_{\mathbb{R}^{N-2}} e^{-\mathcal{H}_{[-1,N]}(\varphi)} \left(\prod_{j=1}^{i-1} d\varphi_j\right) \delta_0(d\varphi_i) d\varphi_{i+1} \prod_{k=i+2}^{N-1} (\varepsilon\delta_0(d\varphi_k) + d\varphi_k)$$

$$\leq C_+ \int_{\mathbb{R}^{N-3}} e^{-\mathcal{H}_{[-1,N]}(\varphi)} \left(\prod_{j=1}^{i-1} d\varphi_j\right) \delta_0(d\varphi_i)\delta_0(d\varphi_{i+1}) \prod_{k=i+2}^{N-1} (\varepsilon\delta_0(d\varphi_k) + d\varphi_k) \quad (5.25)$$

and

$$\int_{\mathbb{R}^{N-2}} e^{-\mathcal{H}_{[-1,N]}(\varphi)} \left(\prod_{j=1}^{i-1} d\varphi_j\right) \delta_0(d\varphi_i) d\varphi_{i+1} \prod_{k=i+2}^{N-1} (\varepsilon\delta_0(d\varphi_k) + d\varphi_k)$$

$$\geq C_- \int_{\mathbb{R}^{N-3}} e^{-\mathcal{H}_{[-1,N]}(\varphi)} \left(\prod_{j=1}^{i-1} d\varphi_j\right) \delta_0(d\varphi_i)\delta_0(d\varphi_{i+1}) \prod_{k=i+2}^{N-1} (\varepsilon\delta_0(d\varphi_k) + d\varphi_k) \ , \quad (5.26)$$

where C_-, C_+ are not depending on N and $i = 1, ..., N-2$.

5.3 SMOOTH PHASE TRANSITION

Although we couldn't prove it, there are reasons why assumption (5.14) should be true, cf. Appendix A.6.

From this assumption it follows easily by Lemma 1.20 for $\varepsilon \leq 1$

$$\mathcal{Z}_{\varepsilon,N} \leq \mathcal{Z}_{0,N} + 2C_+ \varepsilon \sum_{i=1}^{N-1} \mathcal{Z}_{0,i} \mathcal{Z}_{\varepsilon,N-i-1} \qquad (5.27)$$

and

$$\mathcal{Z}_{\varepsilon,N} \geq \mathcal{Z}_{0,N} + C_- \varepsilon \sum_{i=1}^{N-1} \mathcal{Z}_{0,i} \mathcal{Z}_{\varepsilon,N-i-1} , \qquad (5.28)$$

where $\mathcal{Z}_{\varepsilon,0} := 1$, $\mathcal{Z}_{0,0} := 0$ and $\mathcal{Z}_{\varepsilon,1} = 1$. Recall at this point that by Proposition 1.11

$$\mathcal{Z}_{0,N} = \frac{\lambda^{N-1}}{\sqrt{N-1}} \left(\frac{1}{c_2 + o(1)} \right)^{1/2} . \qquad (5.29)$$

and λ is the spectral radius from Proposition 1.5. Then the "renewal-inequalities" in (5.27) and (5.28) lead us to

Lemma 5.15 *Under assumption (5.14) the following bounds hold for the pinning model in $d = 1$*

$$-\log(x_1(\varepsilon)) \leq F(\varepsilon) \leq -\log(x_2(\varepsilon))$$

where $x_i = x_i(\varepsilon) \in (0,1)$ ($i = 1,2$) is the unique solution of

$$g(x_1) := x_1^2 + \sum_{n=2}^{\infty} \frac{x_1^{n+1}}{\sqrt{n-1}} \left(\frac{1}{c_2 + o(1)} \right)^{1/2} = \frac{\lambda^2}{C_- \varepsilon} . \qquad (5.30)$$

and

$$g(x_2) = x_2^2 + \sum_{n=2}^{\infty} \frac{x_2^{n+1}}{\sqrt{n-1}} \left(\frac{1}{c_2 + o(1)} \right)^{1/2} = \frac{\lambda^2}{2C_+ \varepsilon} .$$

Proof Fix any $\varepsilon > 0 = \varepsilon_c$ and take $x := x(\varepsilon) \in (0,1)$ as the corresponding unique solution of $g(x) = \lambda^2/(2C_+ \varepsilon)$. We extend here the idea of the proof in [7] and prove first the upper bound. For $n \geq 1$ we define

$$u_n := \frac{x^n \mathcal{Z}_{\varepsilon,n}}{\lambda^{n-1}} \quad , \quad a_n := \frac{2C_+ \varepsilon x^{n+1} \mathcal{Z}_{0,n}}{\lambda^{n+1}} \quad , \quad b_n := \frac{a_n \lambda^2}{2C_+ \varepsilon x} .$$

We set $a_0 := b_0 := 0$ and $u_0 := \lambda$. In particular $a_1 = 2C_+ \varepsilon x^2 \lambda^{-2}$, $b_1 = x$ and due to (5.29) for $n \geq 2$

$$a_n = \frac{2C_+ \varepsilon}{\lambda^2} \frac{x^{n+1}}{\sqrt{n-1}} \left(\frac{1}{c_2 + o(1)} \right)^{1/2}$$

Because of (5.27) we obtain

$$u_n \geq b_n + \sum_{i=0}^{n-1} a_i u_{n-i-1} = b_n + (a_{-1} u_n + a_0 u_{n-1} + \ldots + a_{n-1} u_0) \quad , \quad a_{-1} := 0$$

$$= b_n + (\tilde{a}_0 u_n + \ldots + \tilde{a}_n u_0) \qquad (5.31)$$

where we set $\tilde{a}_i := a_{i-1}$ for $i = 0, 1, 2, \ldots$. Now let $\tilde{u}_n$ be defined by

$$\tilde{u}_n = b_n + \sum_{i=0}^{n} \tilde{a}_i \tilde{u}_{n-i} \quad , \quad \tilde{u}_0 = \lambda \ . \tag{5.32}$$

By definition of a_n and the choice of $x = x(\varepsilon)$ we know that

$$\sum_{n=0}^{\infty} \tilde{a}_n = \sum_{n=0}^{\infty} a_{n-1} = \frac{2C_+\varepsilon}{\lambda^2} \left(x^2 + \sum_{n=2}^{\infty} \frac{x^{n+1}}{\sqrt{n-1}} \left(\frac{1}{c_2 + o(1)} \right)^{1/2} \right) = \frac{2C_+\varepsilon}{\lambda^2} g(x(\varepsilon)) = 1 \ .$$

But this is enough to apply a theorem from the renewal theory (cf. [15] chap.XIII) on the sequence $\tilde{u}_n$ defined by (5.32) and we get

$$\lim_{n \to \infty} \tilde{u}_n = \frac{B}{A}$$

where

$$B := \sum_{n=0}^{\infty} b_n = x + \sum_{n=2}^{\infty} \frac{x^n}{\sqrt{n-1}} \left(\frac{1}{c_2 + o(1)} \right)^{1/2} = x(\varepsilon) g(x(\varepsilon)) = x(\varepsilon) \frac{\lambda^2}{2C_+\varepsilon}$$

and

$$A := \sum_{n=0}^{\infty} n \tilde{a}_n = \frac{2C_+\varepsilon}{\lambda^2} \left[2x^2 + \sum_{n=2}^{\infty} \frac{(n+1)x^{n+1}}{\sqrt{n-1}} \left(\frac{1}{c_2 + o(1)} \right)^{1/2} \right] = \frac{2C_+\varepsilon}{\lambda^2} x(\varepsilon) g'(x(\varepsilon)) \ .$$

Therefore the limit is

$$\lim_{n \to \infty} \tilde{u}_n = \frac{\lambda^4}{4C_+^2 \varepsilon^2 g'(x)} \tag{5.33}$$

The next what we show is that for every sequence u_n, which fulfills (5.31) and $u_0 = \lambda$, it holds

$$\lim_{n \to \infty} u_n \leq \lim_{n \to \infty} \tilde{u}_n \ . \tag{5.34}$$

Set $\hat{u}_n := \tilde{u}_n - u_n$ for $n \geq 0$, then by (5.31) and (5.32) we have

$$\hat{u}_n \geq \sum_{i=0}^{n} \tilde{a}_i \hat{u}_{n-i} \quad , \quad \hat{u}_0 = 0 \ . \tag{5.35}$$

We claim that $\hat{u}_n \geq 0$ for every $n \in \mathbb{N}_0$. For $n = 0$ it is $\hat{u}_0 = \lambda - \lambda = 0$. Let $\hat{u}_0, \hat{u}_1, \ldots, \hat{u}_n \geq 0$ then by (5.35) and induction it follows

$$\hat{u}_{n+1} \geq \sum_{i=0}^{n+1} \tilde{a}_i \hat{u}_{n+1-i} = \tilde{a}_1 \hat{u}_n + \ldots + \tilde{a}_{n+1} \hat{u}_0 \geq 0$$

due to the fact, that $\tilde{a}_0 = \hat{u}_0 = 0$ and $\tilde{a}_1, \ldots, \tilde{a}_n \geq 0$. So we get $\lim_{n \to \infty} \hat{u}_n \geq 0$ and as we know from (5.33) that $\lim_{n \to \infty} \tilde{u}_n$ exists, we get finally (5.34) and

$$\lim_{n \to \infty} u_n \leq \frac{\lambda^4}{4C_+^2 \varepsilon^2 g'(x)} \ .$$

5.3 Smooth phase transition

Therefore we obtain from the definition of u_n

$$\mathcal{Z}_{\varepsilon,N} \preceq \frac{\lambda^{N+3}}{4C_+^2 \varepsilon^2 x^N g'(x)} \ .$$

Dividing by $\mathcal{Z}_{0,N}$ and using (5.29) we have

$$\frac{\mathcal{Z}_{\varepsilon,N}}{\mathcal{Z}_{0,N}} \preceq \frac{\lambda^4 (c_2 + o(1))^{1/2}}{4C_+^2 \varepsilon^2 g'(x)} \frac{\sqrt{N-1}}{x^N}$$

and therefore we obtain

$$F(\varepsilon) = \lim_{N \to \infty} \frac{1}{N} \log\left(\frac{\mathcal{Z}_{\varepsilon,N}}{\mathcal{Z}_{0,N}}\right) \leq -\log(x) \ , \quad x = x(\varepsilon) \ .$$

The lower bound can be proven in an analogous way and one obtains

$$\frac{\mathcal{Z}_{\varepsilon,N}}{\mathcal{Z}_{0,N}} \succeq \frac{\lambda^4 (c_2 + o(1))^{1/2}}{C_-^2 \varepsilon^2 g'(x)} \frac{\sqrt{N-1}}{x^N} \ ,$$

where x is now the solution of $g(x) = \lambda^2/(C_- \varepsilon)$. $\square$

The free partition function in the weak pinning model for dimension $d = 2$ has the property $\mathcal{Z}_{0,N,2} = (\mathcal{Z}_{0,N,1})^2$, cf chapter 4. Therefore in the same way to the proof of Lemma 5.15 we can show

Lemma 5.16 *Under assumption (5.14) the following bounds hold for the pinning model in $d = 2$*

$$-\log(x_1(\varepsilon)) \leq F_2(\varepsilon) \leq -\log(x_2(\varepsilon))$$

where $x_i = x_i(\varepsilon) \in (0,1)$ ($i = 1, 2$) is the unique solution of

$$g(x_1) := x_1^2 + \sum_{n=2}^{\infty} \frac{x_1^{n+1}}{n-1}\left(\frac{1}{c_2 + o(1)}\right) = \frac{\lambda^4}{C_-^2 \varepsilon} \ . \tag{5.36}$$

and

$$g(x_2) = x_2^2 + \sum_{n=2}^{\infty} \frac{x_2^{n+1}}{n-1}\left(\frac{1}{c_2 + o(1)}\right) = \frac{\lambda^4}{4C_+^2 \varepsilon} \ .$$

By the Remark 5.13 and the last two Lemma we can finally state that our mixed pinning model reveals in $d = 1$ a second and in $d = 2$ an infinite order phase transition. Thus the proof of Theorem 5.2 is completed.

A Appendix

A.1 Zerner's theorem

We recall here an infinite dimensional Perron-Frobenius Theorem, stated in [30]. Alternatively confer also [28] Thm. 6.6, where it is presented as a general version of a classical theorem on kernel operators (Theorem of Jentzsch).

Theorem A.1 (Zerner's Theorem)
Let $E := L^p(\mu)$, where $1 \leq p \leq +\infty$ and (X, Σ, μ) is a σ-finite measure space. Suppose $T \in \mathscr{L}(E)$ is an operator given by a $(\Sigma \times \Sigma)$-measurable kernel $K \geq 0$, satisfying these two assumptions:

(i) Some power of T is compact.

(i) $S \in \Sigma$ and $\mu(S) > 0$, $\mu(X \setminus S) > 0$ implies
$$\int_{X \setminus S} \int_S K(s,t) \, d\mu(s) \, d\mu(t) > 0.$$

Then $r(T) > 0$ is a simple eigenvalue of T with a unique normalized eigenfunction f satisfying $f(s) > 0$ μ-a.e. Moreover, if $K(s,t) > 0$ $(\mu \otimes \mu)$-a.e. then every other eigenvalue λ of T has modulus $|\lambda| < r(T)$.

A.2 Toeplitz Matrices and determinants

In chapter 1 we were interested in handling some determinants in order to prove proposition 1.11. We will first present a method how to compute such determinants and then give the calculations in our case. We will need some definitions.

Definition A.2 *A Toeplitz band matrix of grades (p,q) is a square matrix $[a_{i,j}]$ whose elements depend only on the difference of the indices, i.e. $a_{i,j} = a_{i-j}$ and satisfy $a_{-p} \neq 0$, $a_q \neq 0$ and $a_r = 0$ if $r > q$ or $r < -p$.*

Definition A.3 *We set $\boldsymbol{T} := \{z \in \mathbb{C} \,|\, |z| = 1\}$ and call*

$$a(t) = \sum_{j=-p}^{q} a_j \, t^j \quad (t \in \boldsymbol{T}), \quad p \geq 1, \ q \geq 1, \ a_{-p} a_q \neq 0 \ . \tag{A.1}$$

the symbol of the corresponding Toeplitz band matrix of grades (p,q). Furthermore we denote by $D_n(a)$ the determinant of the corresponding square matrix $T_n(a)$, here n refers to the dimension of the matrix.

For the next theorem confer [8] Thm.5.29

Theorem A.4 *Let a be given by (A.1) and write*

$$a(t) = a_q t^{-p} \prod_{j=1}^{p+q}(t - \varrho_j) \quad (t \in T) .$$

If the zeros $\varrho_1, ..., \varrho_{p+q}$ are pairwise distinct then for every $n \geq 1$,

$$D_n(a) = \sum_M C_M\, w_M^n,$$

where the sum is taken over all $\binom{p+q}{p}$ subsets $M \subset \{1, 2, ..., p+q\}$ of cardinality $|M| = p$ and, with $\overline{M} := \subset \{1, 2, ..., p+q\} \setminus M$,

$$w_M := (-1)^q a_q \prod_{j \in \overline{M}} \varrho_j, \qquad C_M := \prod_{j \in \overline{M}} \varrho_j^p \prod_{\substack{j \in \overline{M} \\ k \in M}} (\varrho_j - \varrho_k)^{-1} .$$

In [8] there is also an useful supplement that if $z^p a(z)$ has multiple zeros, then $D_n(a)$ can be found by first perturbating a and by subsequently passing to an appropriate limit in the formula delivered by Theorem A.4.

Before going into our case, let us first consider two simple cases, the gradient and Laplacian case. The corresponding matrices for these models are A_n and B_n in the proof of Proposition 1.11. The symbols are for A_n:

$$a(t) = 2 - t - \frac{1}{t} = -\frac{1}{t}(t-1)^2$$

with the roots $\{1, 1\}$ and for the Laplacian model

$$a(t) = 6 - 4t + t^2 - \frac{4}{t} + \frac{1}{t^2} = \frac{1}{t}(t-1)^4$$

with the roots $\{1, 1, 1, 1\}$. To apply the Theorem we perturb the roots to $\{1, 1 + \varepsilon\}$ and $\{1, 1 + \varepsilon, 1 + 2\varepsilon, 1 + 3\varepsilon\}$ for some $\varepsilon \in \mathbb{R}$. In order to get a feeling we do the calculation just in the gradient case. Here we have

$$D_n(a) = \lim_{\varepsilon \to 0} (C_1 w_1^n + C_2 w_2^n) = \lim_{\varepsilon \to 0} \left(\frac{1+\varepsilon}{\varepsilon}(1+\varepsilon)^n + \frac{1}{-\varepsilon} 1^n\right)$$
$$= \lim_{\varepsilon \to 0} \frac{1}{\varepsilon}\left((1+\varepsilon)^{n+1} - 1\right) = n + 1 .$$

This fact plays an important role in the investigation of localization in the Gaussian gradient model. Similarly also the determinant for the Laplacian case can be computed

$$\det(B_n) = \frac{1}{12}(2+n)^2(3 + 4n + n^2) .$$

Now we are going to show the result for our model, i.e. equation (1.11). Unfortunately this is quite a mess, because not all roots of the corresponding symbol equal to one. The matrix we are dealing with is $\alpha A_n + \beta B_n$. The symbol equals here

$$a(t) = 2\alpha + 6\beta - (\alpha + 4\beta)t + \beta t^2 - \frac{\alpha + 4\beta}{t} + \frac{\beta}{t^2} \, . \tag{A.2}$$

Furthermore $p = q = 2$ and we have $\binom{p+q}{p} = 6$ subsets $M \subset \{1, 2, ..., 4\}$ of cardinality $|M| = 2$. The matrix elements are

$$(a_{-2}, a_{-1}, a_0, a_1, a_2) = (\beta, -(\alpha + 4\beta), 2\alpha + 6\beta, -(\alpha + 4\beta), \beta) \, .$$

In order to prevent multiple roots we perturb the roots of $a(.)$ in (A.2) as follows:

$$\varrho_1 = 1 \, , \ \varrho_2 = 1 + \varepsilon \, , \ \varrho_3 = \frac{\alpha + 2\beta - \sqrt{\alpha}\sqrt{\alpha + 4\beta}}{2\beta} + 2\varepsilon \, , \ \varrho_4 = \frac{\alpha + 2\beta + \sqrt{\alpha}\sqrt{\alpha + 4\beta}}{2\beta} + 3\varepsilon \, .$$

We have then

$$C_1 = \frac{\varrho_4^2 \varrho_3^2}{(\varrho_3 - 1)(\varrho_3 - \varrho_2)(\varrho_4 - 1)(\varrho_4 - \varrho_2)} \, , \ C_2 = \frac{\varrho_2^2 \varrho_4^2}{(\varrho_2 - \varrho_3)(\varrho_2 - 1)(\varrho_4 - 1)(\varrho_4 - \varrho_3)}$$

$$C_3 = \frac{\varrho_2^2 \varrho_3^2}{(\varrho_2 - \varrho_4)(\varrho_2 - 1)(\varrho_3 - 1)(\varrho_3 - \varrho_4)} \, , \ C_4 = \frac{\varrho_4^2}{(1 - \varrho_2)(1 - \varrho_3)(\varrho_4 - \varrho_2)(\varrho_4 - \varrho_3)}$$

$$C_5 = \frac{\varrho_3^2}{(1 - \varrho_2)(1 - \varrho_4)(\varrho_3 - \varrho_2)(\varrho_3 - \varrho_4)} \, , \ C_6 = \frac{\varrho_2^2}{(1 - \varrho_3)(1 - \varrho_4)(\varrho_2 - \varrho_3)(\varrho_2 - \varrho_4)}$$

and

$$w_1 = \beta \varrho_4 \varrho_3 \, , \ w_2 = \beta \varrho_2 \varrho_4 \, , \ w_3 = \beta \varrho_2 \varrho_3 \, , \ w_4 = \beta \varrho_4 \, , \ w_5 = \beta \varrho_3 \, , \ w_6 = \beta \varrho_2 \, .$$

Finally we have used the computer algebra system "Mathematica" to evaluate the following expression.

$$\det(\alpha A_n + \beta B_n) = \lim_{\varepsilon \to 0} \sum_{k=1}^{6} C_k w_k^n = c_1^{\alpha,\beta} \sigma_+^{n-1} + c_2^{\alpha,\beta} \sigma_+^{n-1}(n-1) + o(\sigma_+^{n-1})$$

with constants which we computed exactly, but we indicate here only the crucial one

$$c_2^{\alpha,\beta} = \frac{2\beta^2 \sqrt{\alpha} + \alpha^2 \sqrt{\alpha + 4\beta} + \alpha^{5/2} + 2\alpha\beta\sqrt{\alpha + 4\beta} + 4\sqrt{\alpha}\alpha\beta}{2\alpha^2 \sqrt{\alpha + 4\beta}} \, .$$

The constant σ_+ was defined in Proposition 1.5.

A.3 Inverse elements of M_n

In the proof of Proposition 1.12 we were dealing with the inverse of a banded Matrix M_n which wasn't of Toeplitz type, due to the free boundary conditions in the corresponding model. The inverse was needed to obtain the means and covariances for the Gaussian density $\varphi_n^{(a,b)}(\cdot, \cdot)$ defined in 1.21. Rózsa gives in [26] a constructive method which is based on the theory of linear difference equations. We are going to use Theorem 4 of [26], but first we define

Definition A.5 *A strict band matrix of grades (p,q) is a square matrix $[a_{i,j}]$ whose elements satisfy $a_{i,j} = 0$ for $j > i + p$ and $j < i - q$. Furthermore it has non-vanishing elements in the p-th diagonal above and in the q-th diagonal below the the main diagonal.*

Apparently M_n is a strictly banded matrix with $p = q = 2$. We enlarge it to an $n \times (n+p+q)$ matrix in such a way that a complete band matrix consisting of $p + q + 1$ lines parallel to the main diagonal results. The matrix determines a system of linear equations consisting of n equations and $n + p + q$ unknowns. Let $r_i(x)$, $i = 1, ..., 4$ be linearly independent solutions of the homogeneous linear difference equation ($x = 0, ..., n - 3$)

$$\beta r(x+4) - (\alpha + 4\beta)r(x+3) + (2\alpha + 6\beta)r(x+2) - (\alpha + 4\beta)r(x+1) + \beta r(x) = 0 \ . \quad \text{(A.3)}$$

This has the characteristic equation

$$\beta \lambda^4 - (\alpha + 4\beta)\lambda^3 + (2\alpha + 6\beta)\lambda^2 - (\alpha + 4\beta)\lambda + \beta = 0$$

with the roots

$$\lambda_1 = 1 \ , \ \lambda_2 = 1 \ , \ \lambda_3 = \frac{\alpha + 2\beta - \sqrt{\alpha}\sqrt{\alpha + 4\beta}}{2\beta} \ , \ \lambda_4 = \frac{\alpha + 2\beta + \sqrt{\alpha}\sqrt{\alpha + 4\beta}}{2\beta} \ .$$

Hence a fundamental system of the difference equation is given as

$$r_1(k) = 1 \ , \ r_2(k) = k \ , \ r_3(k) = \lambda_3^k \ , \ r_4(k) = \lambda_4^k \ , \ k = 0, ..., n - 3 \ . \quad \text{(A.4)}$$

We have to be careful, because M_n is not Toeplitz and some perturbation in the last two rows are present. We set

$$\kappa := \frac{\sqrt{\alpha}\sqrt{\alpha + 4\beta} - \alpha}{2}$$

Therefore additionally to (A.3) there are two equations more. Namely for $x = n - 2$ we have

$$\beta r(n+2) - (\alpha + 2\beta + \kappa)r(n+1) + (2\alpha + 5\beta + \kappa)r(n) - (\alpha + 4\beta)r(n-1) + \beta r(n-2) = 0 \ ,$$

and for $x = n - 1$

$$\beta r(n+3) - (\alpha + 2\beta + \kappa)r(n+2) + (\alpha + \beta + \kappa)r(n+1) - (\alpha + 2\beta + \kappa)r(n) + \beta r(n-1) = 0 \ .$$

Let us define

$$D(x_1, x_2, x_3, x_4) := \det r_i(x_j) = \det \begin{pmatrix} r_1(x_1) & r_1(x_2) & r_1(x_3) & r_1(x_4) \\ r_2(x_1) & r_2(x_2) & r_2(x_3) & r_2(x_4) \\ r_3(x_1) & r_3(x_2) & r_3(x_3) & r_3(x_4) \\ r_4(x_1) & r_4(x_2) & r_4(x_3) & r_4(x_4) \end{pmatrix}$$

and for the so called supporting points $\{0, 1, n + 2, n + 3\}$

$$g_1(x) := \frac{D(x, 1, n+2, n+3)}{D(0, 1, n+2, n+3)} \quad , \quad g_2(x) := \frac{D(0, x, n+2, n+3)}{D(0, 1, n+2, n+3)}$$

$$h_1(x) := \frac{D(0, 1, n+2, x)}{D(0, 1, n+2, n+3)} \quad , \quad h_2(x) := \frac{D(0, 1, x, n+3)}{D(0, 1, n+2, n+3)}$$

$$s_1(y) := \frac{D(y, y+1, y+2, 0)}{D(y, y+1, y+2, y+3)} \quad , \quad s_2(y) := \frac{D(y, y+1, y+2, 1)}{D(y, y+1, y+2, y+3)}$$

$$t_1(y) := \frac{D(y, y+1, y+2, n+3)}{D(y, y+1, y+2, y+3)} \quad , \quad t_2(y) := \frac{D(y, y+1, y+2, n+2)}{D(y, y+1, y+2, y+3)} \ .$$

Now to obtain the values for the fundamental system also for $n+2$ and $n+3$ we compute them recursively from above equations for $x = n-2$ and $n-1$ for each $r_i(\cdot)$. If we set $(M_n)^{-1} =: W = (w_{i,j})$, $i,j = 1,...,n$ and take the boundary conditions

$$f(0) = f(1) = f(n+2) + f(n+3) = 0$$

in the inhomogeneous linear difference equation then Theorem 4 of [26] tells us that

$$w_{i,j} = \begin{cases} g_1(i+1)s_1(j) + g_2(i+1)s_2(j) & \text{, if } i > j-2 \\ -h_1(i+1)t_1(j) - h_2(i+1)t_2(j) & \text{, if } i < j+2 \end{cases}.$$

Now the quantities needed for the proof of Proposition 1.12 can be computed from (1.19) and (1.20) as follows

$$\mu_j = w_{j,1}(\alpha b + \beta(3b+a)) - w_{j,2}\beta b \quad, j = 1,2, n-1, n$$

and

$$\Upsilon = \alpha b^2 + \beta[(b+a)(3b+a) - b(b+2a)] - (\mu_1(\alpha b + \beta(3b+a)) - \mu_2 \beta b) .$$

The results is stated in Proposition 1.12. Here we have used again "Mathematica" for those calculations.

A.4 Recursive representation of $\{Y_i\}_{i \in \mathbb{Z}}$

We are going to show here the representation (3.4), i.e. the Markov chain $\{Y_i\}_{i \in \mathbb{Z}^+}$ constructed in subsection 1.4.1 equals under $\mathrm{P}^{(a,b)}$:

$$\tilde{Y}_n := \gamma^n a + \gamma^{n-1}\varepsilon_1 + ... + \gamma^0 \varepsilon_n .$$

First of all clearly $\tilde{Y}_0 = a = Y_0$. Now completing the squares in the transition probability (1.4) we obtain

$$\mathrm{P}^{(a,b)}(Y_{n+1} = dy | Y_n = x) \sim \mathcal{N}(\gamma x, \sigma^2)$$

with (sgn denotes the signum function):

$$\gamma = \left(\frac{\alpha + 2\beta - \sqrt{\alpha}\sqrt{\alpha + 4\beta}}{\alpha + 2\beta + \sqrt{\alpha}\sqrt{\alpha + 4\beta}}\right)^{1/2} \cdot \mathrm{sgn}(\beta) \text{ and } \sigma^2 = \frac{1}{\sigma_+} = \frac{2}{\alpha + 2\beta + \sqrt{\alpha}\sqrt{\alpha + 4\beta}} .$$

Now let $\{\varepsilon_i\}_{i \in \mathbb{Z}^+}$ be a sequence of i.i.d. centered Gaussian variables $\sim \mathcal{N}(0, \sigma^2)$. The following recursive structure of $\{\tilde{Y}_i\}_{i \in \mathbb{Z}^+}$ can be discovered

$$\tilde{Y}_n = \gamma \tilde{Y}_{n-1} + \varepsilon_n ,$$

since with that indeed

$$\gamma \tilde{Y}_{n-1} + \varepsilon_n = \gamma^2 \tilde{Y}_{n-2} + \gamma \varepsilon_{n-1} + \varepsilon_n = ... = \gamma^n a + \gamma^{n-1} \varepsilon_1 + ... + \gamma^0 \varepsilon_n = \tilde{Y}_n .$$

Finally due to independence of ε_{n+1} and $\tilde{Y}_n$ we obtain the same transition probability as above

$$\mathrm{P}^{(a,b)}(\tilde{Y}_{n+1} = dy|\tilde{Y}_n = x) = \mathrm{P}^{(a,b)}(\varepsilon_{n+1} = dy - \gamma\tilde{Y}_n|\tilde{Y}_n = x) \sim \mathcal{N}(\gamma x, \sigma^2) \ .$$

What appears here not to be too difficult to see, was a long time hidden for us. Originally we came across the recursive representation just by chance, when we were interested in the n-th step transition probability of our Markov chain defined in subsection 1.4.1. Namely, if we define

$$P(x, dy) := \mathrm{P}^{(a,b)}(Y_{n+1} = dy|Y_n = x)$$

and the n-th step transition probability by

$$P^n(x, B) := \int_{\mathbb{R}} P^{n-1}(x, dy) P(y, B) \quad , \ x \in \mathbb{R}, B \in \mathcal{B}(\mathbb{R}), \ n \geq 1$$

$$P^0(x, .) := \mathrm{P}^{(x,b)} \circ Y_0^{-1}(.) = \delta_x(.) \ ,$$

then we have calculated

$$P^n(x, .) = \mathrm{P}^{(x,b)}(Y_n \in .) \sim \mathcal{N}\left(\gamma^n x, \frac{1 - \gamma^{2n}}{1 - \gamma^2}\sigma^2\right) \ .$$

Looking for a class of Markov chains that satisfy similar properties, we have found out that $\{Y_i\}_{i \in \mathbb{Z}^+}$ is nothing else as an autoregressive process of order one. But such processes are defined recursively and so we got to the representation (3.4).

Remark A.6
We didn't embed the case $\beta < 0$ and at the same time $\alpha + 4\beta \geq 0$ in the thesis for the following reason. These cases are treated in the same way as for $\alpha, \beta > 0$ and we obtain the same results for the localization behavior. This holds also in the extremal case $\alpha = -4\beta$. Let us make a short comment on this, since in the last case $\gamma = -1$ and one could think that something different happens here. The crucial comment is about the behavior of the density $\varphi_n^{(-x,0)}(y, 0)$. Recall the representation (1.21) for $n \geq 2$. For $\alpha = -4\beta$ we have calculated

$$\tilde{f}_{x,y}(n) = \varphi_n^{(-x,0)}(y, 0)\mathbb{1}_{\{y \neq 0\}}$$

$$= \frac{1}{2\pi\sqrt{\det(\Sigma_n)}} \exp\left\{-\frac{1}{2}\langle\begin{pmatrix}y - \mu_{n-1}^{\alpha,\beta}(-x,0) \\ -\mu_n^{\alpha,\beta}(-x,0)\end{pmatrix}, \Sigma_n^{-1}\begin{pmatrix}y - \mu_{n-1}^{\alpha,\beta}(-x,0) \\ -\mu_n^{\alpha,\beta}(-x,0)\end{pmatrix}\rangle\right\}\mathbb{1}_{\{y \neq 0\}}$$

$$= \frac{1}{2\pi\sqrt{\det(\Sigma_n)}} \exp\left\{-\frac{1}{2}\left(x^2 a_n + y^2 b_n + 2xy c_n\right)\right\}$$

where

$$\Sigma_n = \begin{pmatrix} \mathrm{Cov}(W_{n-1}, W_{n-1}) & \mathrm{Cov}(W_{n-1}, W_n) \\ \mathrm{Cov}(W_n, W_{n-1}) & \mathrm{Cov}(W_n, W_n) \end{pmatrix} \ ,$$

$$\mu_n^{\alpha,\beta}(-x, 0) = E_{\mathrm{P}^{(-x,0)}} W_n = -nx$$

and

$$a_n = \frac{2(6 + n(2n - 5))}{n(n+1)(2+n)\sigma^2} \quad , \quad b_n = \frac{2(3 + 2n)}{(n + n^2)\sigma^2} \quad , \quad c_n = \frac{2(n - 3)}{(n + n^2)\sigma^2} \ .$$

Furthermore it is
$$\frac{1}{\det \Sigma_n} = \frac{12}{n(n+1)^2(2+n)\sigma^2},$$
so a behavior like in the Laplacian case, confer subsection 4.4.4. The crucial difference here is that in the exponent we have the asymptotical behavior x^2/n, whereas in the Laplacian case there is only x^2. We don't want to go further into detail, but this is the reason why the localization is the same like in chapter 1 (cf. also Lemma 1.16).

A.5 The conditional processes

Concerning Proposition 3.7 and Proposition 3.8.
In this part we show the representations of the conditional processes in Proposition 3.7 and Proposition 3.8. We will see that there is an explicit representation of the conditional integrated Markov chain (under $P^{(-x,0)}$), namely

$$\widehat{W}_n(y, N) = W_n - (W_N - y)r_1(n) - W_{N+1}r_2(n) \,, \; n = 1, ..., N-1$$

where
$$r_1(n) = \frac{s_1(n)}{r(n)} \quad \text{and} \quad r_2(n) = \frac{s_2(n)}{r(n)}$$

and

$$r(n) = (-1+\gamma)\left(-1+\gamma^{1+N}\right)\left(-N+\gamma\left(2+N+\gamma^N(-2+(-1+\gamma)N)\right)\right),$$
$$s_1(n) = (-n+\gamma(1-\gamma^n+n)) + \gamma^{3+2N-n}(1+\gamma^n(-1+(-1+\gamma)n))$$
$$+ \gamma^{N-n}\left(\gamma^n\left(-\gamma+\gamma^3\right)(1-n+N) + \gamma^{2+2n}(2+N-\gamma(1+N)) + \gamma(1+N-\gamma(2+N))\right),$$
$$s_2(n) = \gamma\left(\gamma^{1+n}+n-\gamma(1+n)\right) + \gamma^{2+2N-n}(-1+\gamma^n(1+n-\gamma n))$$
$$+ \gamma^{1+N-n}\left(\gamma+\gamma^n\left(-1+\gamma^2\right)(n-N) - N + \gamma N + \gamma^{1+2n}(-1+(-1+\gamma)N)\right).$$

As we already mentioned in Remark 3.9, the proof of both Propositions is based on the representation of the conditional distributions of $(Y_n | W_N = y, W_{N+1} = 0)$ and $(S_n | S_N = y)$, respectively. These are Gaussian processes and so it is possible to calculate the conditional distributions. In the following we will concentrate on the Markov chain $\{Y_i\}_{i \in \mathbb{Z}^+}$ with its nice representation (3.4).
Let
$$\mu = \begin{bmatrix} \mu_1 \\ \mu_2 \end{bmatrix} \text{ be of size } \begin{bmatrix} N-1 \\ 2 \end{bmatrix}$$

and $\mu_1 := (\mathbb{E}_{P^{(-x,0)}} Y_1, ..., \mathbb{E}_{P^{(-x,0)}} Y_{N-1})^T$, $\mu_2 := (\mathbb{E}_{P^{(-x,0)}} W_N, \mathbb{E}_{P^{(-x,0)}} W_{N+1})^T$. Moreover let
$$\Sigma = \begin{bmatrix} \Sigma_1 & \Sigma_2 \\ \Sigma_3 & \Sigma_4 \end{bmatrix} \text{ be of size } \begin{bmatrix} (N-1)\times(N-1) & (N-1)\times 2 \\ 2\times(N-1) & 2\times 2 \end{bmatrix}$$

and $\Sigma_1 := \{\text{Cov}(Y_i, Y_j)\}_{i,j=1,...,N-1}$, $\Sigma_2 := \{\text{Cov}(Y_i, W_j)\}_{i=1,...,N-1, j=1,2}$,
$\Sigma_3 := \{\text{Cov}(W_i, Y_j)\}_{i=1,2, j=1,...,N-1}$, $\Sigma_4 := \{\text{Cov}(W_i, W_j)\}_{i=1,2, j=1,2}$. Then it is well

known that
$$\mathscr{L}\left(((Y_i)_{i=1,\ldots,N-1} \mid W_N = y, W_{N+1} = 0)\right) \sim \mathcal{N}\left(\bar{\mu}, \bar{\Sigma}\right),$$
where
$$\bar{\mu} = \mu_1 + \Sigma_2 \Sigma_4^{-1}\left(\begin{bmatrix} y \\ 0 \end{bmatrix} - \mu_2\right) \quad \text{and} \quad \bar{\Sigma} = \Sigma_1 - \Sigma_2 \Sigma_4^{-1} \Sigma_3 \ .$$

Remark A.7
The matrix $\bar{\Sigma}$ is called the Schur complement of Σ_4 in Σ. Observe that knowing that $W_N = y, W_{N+1} = 0$ changes the variance, although the new variance does not depend on the specific values $(y, 0)$. Instead we just have to shift the mean by $\Sigma_2 \Sigma_4^{-1}\left([y\ 0]^T - \mu_2\right)$.

In greater detail we can write for an $i \in \{1, \ldots, N-1\}$
$$\bar{\mu}_i = \mathbb{E}_{\mathrm{P}(-x,0)} Y_i + \mathrm{Cov}(Y_i, W_N) y \left(\Sigma_4^{-1}\right)_{1,1} + \mathrm{Cov}(Y_i, W_{N+1}) y \left(\Sigma_4^{-1}\right)_{2,1}$$
$$- \left[\mathrm{Cov}(Y_i, W_N)\left(\left(\Sigma_4^{-1}\right)_{1,1} \mathbb{E}_{\mathrm{P}(-x,0)} W_N + \left(\Sigma_4^{-1}\right)_{1,2} \mathbb{E}_{\mathrm{P}(-x,0)} W_{N+1}\right)\right.$$
$$\left. + \mathrm{Cov}(Y_i, W_{N+1})\left(\left(\Sigma_4^{-1}\right)_{2,1} \mathbb{E}_{\mathrm{P}(-x,0)} W_N + \left(\Sigma_4^{-1}\right)_{2,2} \mathbb{E}_{\mathrm{P}(-x,0)} W_{N+1}\right)\right]$$

and for $1 \leq i \leq j \leq N-1$
$$\bar{\Sigma}_{i,j} = \mathrm{Cov}(Y_i, Y_j) - \left[\mathrm{Cov}(Y_i, W_N)\left(\left(\Sigma_4^{-1}\right)_{1,1} \mathrm{Cov}(W_N, Y_j) + \left(\Sigma_4^{-1}\right)_{1,2} \mathrm{Cov}(W_{N+1}, Y_j)\right)\right.$$
$$\left. + \mathrm{Cov}(Y_i, W_{N+1})\left(\left(\Sigma_4^{-1}\right)_{2,1} \mathrm{Cov}(W_N, Y_j) + \left(\Sigma_4^{-1}\right)_{2,2} \mathrm{Cov}(W_{N+1}, Y_j)\right)\right] \ .$$

With some effort on calculations we obtain
$$\bar{\mu}_i = \mathbb{E}_{\mathrm{P}(-x,0)}[Y_i | W_N = y, W_{N+1} = 0] = -\gamma^i x + a_1(n) x \gamma \frac{1-\gamma^N}{1-\gamma} + a_2(n) x \gamma \frac{1-\gamma^{N+1}}{1-\gamma} + y a_1(n) \ ,$$
where
$$a_1(n) = \frac{\gamma^{-i}(-1+\gamma^i)\left(-\gamma^i + \gamma^{3+2N} + \gamma^{1+N}(1+N-\gamma(2+N) + \gamma^i(2+N-\gamma(1+N)))\right)}{(-1+\gamma^{1+N})(-N+\gamma(2+N+\gamma^N(-2+(-1+\gamma)N)))}$$
$$a_2(n) = \frac{\gamma^{1-n}(-1+\gamma^n)(\gamma^n - \gamma^{1+2N} + \gamma^N(\gamma - N + \gamma N + \gamma^n(-1+(-1+\gamma)N)))}{(-1+\gamma^{1+N})(-N+\gamma(2+N+\gamma^N(-2+(-1+\gamma)N)))} \ .$$

Moreover we have the following covariances for $1 \leq i \leq j \leq N-1$
$$\mathrm{Cov}_{\mathrm{P}(-x,0)}\left((Y_i | W_N = y, W_{N+1} = 0), (Y_j | W_N = y, W_{N+1} = 0)\right)$$
$$= -\frac{1}{(-1+\gamma^2)(-1+\gamma^{1+N})(-N+\gamma(2+N+\gamma^N(-2+(-1+\gamma)N)))} \frac{(1-\gamma)^2}{\sigma^2}$$
$$\cdot \left[\gamma^{-i-j}(-1+\gamma^j)(\gamma^i - \gamma^{1+N})(\gamma^j + \gamma^{1+j} - \gamma^{1+i+N} - \gamma^{2+i+N} - \gamma^i N + \gamma^{2+j+N} N \right.$$
$$- \gamma^{i+j}(1+N) + \gamma^{1+i+j}(1+N) - \gamma^{1+N}(1+N) + \gamma^{2+N}(1+N) + \gamma^{1+i}(2+N)$$
$$\left. - \gamma^{1+j+N}(2+N))\right] \ .$$

A.5 THE CONDITIONAL PROCESSES

The most interesting question now is, whether there is such an explicit representation for the conditional Markov chain like for the Markov chain itself in (3.4). For this purpose let us set for $1 \leq i \leq N-1$

$$\widehat{Y}_i := Y_i - a_1(i)(W_N - y) - a_2(i)W_{N+1} \ . \tag{A.5}$$

In order to check that under $P^{(-x,0)}$

$$\mathscr{L}\left((\widehat{Y}_1, ..., \widehat{Y}_{N-1})\right) = \mathscr{L}\left(((Y_i)_{i=1,...,N-1} \mid W_N = y, W_{N+1} = 0)\right) \ ,$$

all we have to do is to compare the expectations and covariances above to these of $\{\widehat{Y}_i\}_{i=1,...,N-1}$. Indeed, this can be done successfully by some calculations. It was not really obvious why (A.5) should be the right explicit representation and in fact we have spent some time to find it.

Remark A.8 *In the similar manner as before, one can also show that $\{\widehat{Y}_i\}_{i=1,...,N-1}$ under $P^{(-x,0)}$ equals*

$$\left\{Y_i - \gamma^i x - a_1(i)\left(W_N - x\gamma \frac{1-\gamma^N}{1-\gamma} - y\right) - a_2(i)\left(W_{N+1} - x\gamma \frac{1-\gamma^{N+1}}{1-\gamma}\right)\right\}_{i=1,...,N-1}$$

under $P^{(0,0)}$.

To obtain now the explicit representation in Proposition 3.7 we have to sum $\widehat{Y}_i$ in (A.5). This leads for $1 \leq n \leq N-1$ to

$$\widehat{W}_n = (W_n | W_N = y, W_{N+1} = 0) = \sum_{i=1}^n \widehat{Y}_i = W_n - (W_N - y)r_1(n) - W_{N+1}r_2(n) \ ,$$

where

$$r_1(n) = \sum_{i=1}^n a_1(i) \quad \text{and} \quad r_2(n) = \sum_{i=1}^n a_2(i)$$

denote exactly the expressions given in the beginning of Appendix A.5.

The representation in Proposition 3.8 is shown in the same manner. However this is much more simpler, since under $P^{(0,0)}$ one gets

$$\mathscr{L}\left(((S_i)_{i=1,...,N-1} \mid S_N = y)\right) \sim \mathcal{N}\left(\tilde{\mu}, \tilde{\Sigma}\right) \ ,$$

where for $1 \leq i \leq j \leq N-1$

$$\tilde{\mu}_i = \frac{i}{N}y \quad \text{and} \quad \tilde{\Sigma}_{i,j} = \frac{i(N-j)}{N} \ .$$

It is now easy to compare these expectations and covariances to those of

$$\widehat{S}_i := S_i - \frac{i}{N}(S_N - y)$$

Concerning Lemma 3.10

Here we will show that for sufficiently large N there can be found an $q(\gamma, n, N) > 0$, s.t. (3.11) holds. First of all observe that by the last part in this Appendix and Remark A.8 the equality (3.11) is indeed true with

$$q(\gamma, n, N) := \gamma \frac{1-\gamma^n}{1-\gamma} - r_1(n)\gamma \frac{1-\gamma^N}{1-\gamma} - r_2(n)\gamma \frac{1-\gamma^{N+1}}{1-\gamma}.$$

So we are left with the problem of positiveness of this expression for sufficiently large N. We can rewrite q in a more tractable way, i.e.

$$q(\gamma, n, N) = \gamma \frac{1-\gamma^N}{1-\gamma} \left[\frac{1-\gamma^n}{1-\gamma^N} - r_1(n) - r_2(n) - r_2(n) \frac{\gamma^N - \gamma^{N+1}}{1-\gamma^N} \right].$$

Let us set $\tilde{q}(\gamma, n, N) := [\ldots]$ for the brackets above, then by (3.18) and Lemma A.10 we can state

$$\lim_{N\to\infty} \inf_{n=1,\ldots,N-1} \tilde{q}(\gamma, n, N) = \lim_{N\to\infty} \inf_{n=1,\ldots,N-1} \left[1 - \gamma^n - \frac{n}{N} - \frac{\gamma(1-\gamma^n) + \gamma^{N-n+1}}{N(-1+\gamma)} \right]$$

$$= \lim_{N\to\infty} \inf_{n=1,\ldots,N-1} \left[1 - \gamma^n - \frac{n}{N} \right].$$

However it is easily calculated that the infimum is attained in $n = N - 1$:

$$\inf_{n=1,\ldots,N-1} \left[1 - \gamma^n - \frac{n}{N} \right] = 1 - \gamma^{N-1} - \frac{N-1}{N} = -\gamma^{N-1} + \frac{1}{N},$$

which is positive for sufficiently large N (depending on γ). Therefore we get the following asymptotical behavior

$$q(\gamma, n, N) \succeq \frac{\gamma}{1-\gamma} \left(\frac{1}{N} - \gamma^{N-1} \right)$$

and therefore there exists an $N_0(\gamma) \in \mathbb{N}$, s.t. for all $N \geq N_0(\gamma)$ the quantity $q(\gamma, n, N)$ is positive.

A.6 The assumption 5.14

To obtain the order of phase transition for our pinning model in $d = 1$ and $d = 2$, we needed assumption 5.14. Why should it be true? Well, if we consider (5.25) and (5.26) without further pinning after w_{i+1}, then we obtain by the proof of Lemma 3.21 on the one

A.6 The assumption 5.14

hand:

$$\int_{\mathbb{R}^{N-1}} e^{-\mathcal{H}_{[-1,N+1]}(w)} \, dw_1 \cdots dw_{i-1} \delta_0(dw_i) dw_{i+1} \cdots dw_{N-1}$$
$$= \int_{\mathbb{R}} \varphi_{i+1}^{(0,0)}(0, w_{i+1}) \frac{\lambda^{i+1}\nu(0)}{\nu(w_{i+1})} \varphi_{N-i}^{(w_{i+1}, w_{i+1})}(0,0) \frac{\lambda^{N-i}\nu(w_{i+1})}{\nu(0)} \, dw_{i+1}$$
$$= \lambda^{N+1} \int_{\mathbb{R}} \varphi_{i+1}^{(0,0)}(0, w_{i+1}) \varphi_{N-i}^{(w_{i+1}, w_{i+1})}(0,0) \, dw_{i+1} \tag{A.6}$$
$$= \frac{\lambda^{N+1}}{(2\pi)^2 \sqrt{\det \Sigma_{i+1}} \sqrt{\det \Sigma_{N-i}}} \int_{\mathbb{R}} e^{-\frac{1}{2}w_{i+1}^2 (\Sigma_{i+1}^{-1})_{2,2}} e^{-\frac{1}{2}w_{i+1}^2 c(\gamma, N-i)} \, dw_{i+1} \, ,$$

where there exist constants c_1, c_2, such that for all $n \in \mathbb{N}$:

$$0 < c_1 \le (\Sigma_n^{-1})_{2,2} \le c_2 < \infty \quad \text{and} \quad 0 < c(\gamma, n) \le c_2 < \infty \, .$$

On the other hand:

$$\int_{\mathbb{R}^{N-1}} e^{-\mathcal{H}_{[-1,N+1]}(w)} \, dw_1 \cdots dw_{i-1} \delta_0(dw_i) \delta_0(dw_{i+1}) dw_{i+2} \cdots dw_{N-1}$$
$$= \mathcal{Z}_{0,i} \, \mathcal{Z}_{0,N-i-1} = \lambda^{i+1} \lambda^{N-i} \varphi_{i+1}^{(0,0)}(0,0) \, \varphi_{N-i}^{(0,0)}(0,0) \tag{A.7}$$
$$= \frac{\lambda^{N+1}}{(2\pi)^2 \sqrt{\det \Sigma_{i+1}} \sqrt{\det \Sigma_{N-i}}} \, .$$

This means of course that in this case we can substitute dw_{i+1} by $\delta_0(dw_{i+1})$ by paying just universal constants $C, C+$, as was stated in assumption 5.14. Now what happens if we consider

$$\int_{\mathbb{R}^{N-1}} e^{-\mathcal{H}_{[-1,N+1]}(w)} \, dw_1 \cdots dw_{i-1} \delta_0(dw_i) dw_{i+1} \prod_{k=i+2}^{N-1} (\varepsilon \delta_0(dw_k) + dw_k) \, .$$

Of course we can expand as usual for $M = \{i+2, ..., N-1\}$

$$\prod_{k=i+2}^{N-1} (\varepsilon \delta_0(dw_i) + dw_i) = \sum_{A \subseteq M} \varepsilon^{|A|} \left(\prod_{m \in A} \delta_0(dw_m) \right) \left(\prod_{n \in M \setminus A} dw_n \right) \, .$$

Then for a partition $0 < t_1 < \cdots < t_{k-1} < t_k := N$ and $A := \{t_1, ..., t_{k-1}\}$ compare

$$\varepsilon^{k-1} \int_{\mathbb{R}^{N-1}} e^{-\mathcal{H}_{[-1,N+1]}(w)} \, dw_1 \cdots dw_{i-1} \delta_0(dw_i) dw_{i+1} \left(\prod_{m \in A} \delta_0(dw_m) \right) \left(\prod_{n \in M \setminus A} dw_n \right) \tag{A.8}$$

and

$$\varepsilon^{k-1} \int_{\mathbb{R}^{N-1}} e^{-\mathcal{H}_{[-1,N+1]}(w)} \, dw_1 \cdots dw_{i-1} \delta_0(dw_i) \delta_0(dw_{i+1}) \left(\prod_{m \in A} \delta_0(dw_m) \right) \left(\prod_{n \in M \setminus A} dw_n \right) \, . \tag{A.9}$$

Now equation (A.8) can be written as follows

$$\varepsilon^{k-1} \widetilde{F}_{0,dw_{i-1}}(i) \widetilde{F}_{w_{i-1},dy_1}(t_1 - i) \widetilde{F}_{y_1,dy_2}(t_2 - t_1) \cdots \widetilde{F}_{y_{k-1},dy_k}(N - t_{k-1}) \widetilde{F}_{y_k,\{0\}}(1)$$

where for $n \geq 2$
$$\widetilde{F}_{x,dy}(n) = \varphi_n^{(-x,0)}(y,0) \mathbf{1}_{\{y \neq 0\}} \mu(dy) .$$
Whereas equation (A.9) is

$$\varepsilon^{k-1} \widetilde{F}_{0,dw_{i-1}}(i) \widetilde{F}_{w_{i-1},\{0\}}(1) \widetilde{F}_{0,dy_1}(t_1-(i+1)) \widetilde{F}_{y_1,dy_2}(t_2-t_1) \cdots \widetilde{F}_{y_{k-1},dy_k}(N-t_{k-1}) \widetilde{F}_{y_k,\{0\}}(1)$$
$$= \varepsilon^{k-1} \mathcal{Z}_{0,i} \widetilde{F}_{0,dy_1}(t_1-(i+1)) \widetilde{F}_{y_1,dy_2}(t_2-t_1) \cdots \widetilde{F}_{y_{k-1},dy_k}(N-t_{k-1}) \widetilde{F}_{y_k,\{0\}}(1) .$$

Now similarly to equation (A.6) and (A.7) we could consider (now until the first pinning in t_1):

$$\lambda^{t_1+1} \int_{\mathbb{R}} \varphi_{i+1}^{(0,0)}(0, w_{i+1}) \varphi_{t_1-i}^{(w_{i+1}, w_{i+1})}(y_1, 0) \, dw_{i+1} \tag{A.10}$$
$$= \lambda^{t_1+1} \varphi_{i+1}^{(0,0)}(0,0) \varphi_{t_1-i}^{(0,0)}(y_1,0) \int_{\mathbb{R}} e^{-\frac{1}{2} w_{i+1}^2 (\Sigma_{i+1}^{-1})_{2,2}} e^{-\frac{1}{2}(w_{i+1}^2 a_1(\gamma, t_1-i) - w_{i+1} y_1 a_2(\gamma, t_1-i))} \, dw_{i+1}$$

and
$$\lambda^{t_1+1} \varphi_{i+1}^{(0,0)}(0,0) \, \varphi_{t_1-i}^{(0,0)}(y_1,0) . \tag{A.11}$$

Here again there exist constants c_1, c_2, such that for all $n \in \mathbb{N}$:
$$0 < c_1 \leq (\Sigma_n^{-1})_{2,2} , \ a_1(\gamma, n) , \ a_2(\gamma, n) \leq c_2 < \infty .$$

Now clearly the comparison of (A.10) and (A.11) would be straight forward if the integral
$$\int_{\mathbb{R}} e^{-\frac{1}{2} w_{i+1}^2 (\Sigma_{i+1}^{-1})_{2,2}} e^{-\frac{1}{2}(w_{i+1}^2 a_1(\gamma, t_1-i) - w_{i+1} y_1 a_2(\gamma, t_1-i))} \, dw_{i+1} =: I_1(w_{i+1}, y_1)$$

was bounded. The problem is of course the linear dependence on y_1 in the exponent. In this way, by (A.10), this problem propagates to the term with the next contact:

$$\lambda^{t_2+1} \varphi_{i+1}^{(0,0)}(0,0) \, \varphi_{t_1-i}^{(0,0)}(y_1,0) I_1(w_{i+1}, y_1) \, \varphi_{t_2-t_1}^{(-y_1,0)}(y_2,0)$$

and so on to the last one until we reach the right boundary. If we assume in the last equation that there is no further pinning after t_1, then $y_2 = 0$ and the last term would be quadratic in y_1. One could imagine that $I_1(w_{i+1}, y_1)$ can be again "pulled" out by paying constants. However it is not clear how to proceed and there is hope that coming to the right boundary one can always pull out the integral $I_1(w_{i+1}, y_1)$ by paying universally $C_-, C+$.

A.7 Some facts

Calculation A.9 *Denote by $h(\alpha, \beta) := 2\beta + \alpha - \sqrt{\alpha}\sqrt{\alpha + 4\beta}$. We show that this function is strictly positive under assumption (AP) in chapter 1. For this purpose we will just check the derivatives, i.e.*
$$\frac{\partial h}{\partial \beta}(\alpha, \beta) = 2 - \frac{2\sqrt{\alpha}}{\sqrt{\alpha + 4\beta}} \stackrel{!}{=} 0 \iff \beta = 0$$

and
$$\frac{\partial h}{\partial \alpha}(\alpha, \beta) = 1 - \frac{\sqrt{\alpha}}{2\sqrt{\alpha + 4\beta}} - \frac{\sqrt{\alpha + 4\beta}}{2\sqrt{\alpha}} \stackrel{!}{=} 0 \iff \beta = 0 .$$

A.7 Some facts

Moreover, computing the Hessian matrix we obtain

$$\det D_i D_j h(\alpha,\beta)\Big|_{(\alpha,0)} = \frac{16}{\alpha^2} \quad \text{and} \quad \frac{\partial^2 h}{\partial \alpha^2}(\alpha,\beta)\Big|_{(\alpha,0)} = \frac{4}{\alpha},$$

which is enough for the desired positiveness.

Lemma A.10 *For $i = 1, 2$ we have the following limiting behavior*

$$-\frac{1}{1-\gamma} \leq \lim_{N\to\infty} \sup_{n=1,\ldots,N-1} r_i(n) \leq \frac{1}{1-\gamma}.$$

Proof We make use of the explicit representations of r_i in the beginning of Appendix A.5. It can be seen that for $N \to \infty$

$$\sup_{n=1,\ldots,N-1} r_1(n) \sim \sup_{n=1,\ldots,N-1} \left[\frac{-n + \gamma(1 - \gamma^n + n) + \gamma^{N-n+1}(1 + N - \gamma(2+N))}{(1-\gamma)(2\gamma + N(\gamma-1))}\right]$$

$$\sim \sup_{n=1,\ldots,N-1} \left[\frac{-n/N + \gamma n/N + \gamma^{N-n+1}(1-\gamma)}{(1-\gamma)(\gamma-1)}\right]$$

$$\sim \frac{1}{1-\gamma} \sup_{n=1,\ldots,N-1} \left[\frac{n}{N} - \gamma^{N-n+1}\right]$$

and analogously

$$\sup_{n=1,\ldots,N-1} r_2(n) \sim \sup_{n=1,\ldots,N-1} \left[\frac{\gamma(\gamma^{1+n} + n - \gamma(1+n)) + \gamma^{2+N-n} + \gamma^{1+N-n}N(\gamma-1)}{(1-\gamma)(2\gamma + N(\gamma-1))}\right]$$

$$\sim \sup_{n=1,\ldots,N-1} \left[\frac{\gamma(n/N - \gamma n/N) + \gamma^{1+N-n}(\gamma-1)}{(1-\gamma)(\gamma-1)}\right]$$

$$\sim \frac{\gamma}{1-\gamma} \sup_{n=1,\ldots,N-1} \left[\gamma^{N-n} - \frac{n}{N}\right].$$

Therfore we can conclude that

$$\left|\lim_{N\to\infty} \sup_{n=1,\ldots,N-1} r_i(n)\right| \leq \frac{1}{1-\gamma}.$$

□

Formula A.11
In chapter 4 we have used the following general formulas. Let $a > 0$ then

$$\int_{-\infty}^{\infty} \exp\left\{-\frac{1}{2}\langle \begin{pmatrix} y-m_1 \\ -m_2 \end{pmatrix}, \begin{pmatrix} a & b \\ b & d \end{pmatrix} \begin{pmatrix} y-m_1 \\ -m_2 \end{pmatrix}\rangle\right\} dy = \frac{\sqrt{2\pi}}{\sqrt{a}} \exp\left\{-\frac{ad-b^2}{2a}m_2^2\right\}.$$

Moreover, one has

$$\int_r^{\infty} \exp\left\{-\frac{1}{2}\langle \begin{pmatrix} y-m_1 \\ -m_2 \end{pmatrix}, \begin{pmatrix} a & b \\ b & d \end{pmatrix} \begin{pmatrix} y-m_1 \\ -m_2 \end{pmatrix}\rangle\right\} dy$$

$$= \frac{\sqrt{2\pi}}{\sqrt{a}} \exp\left\{-\frac{ad-b^2}{2a}m_2^2\right\} \left(1 + \text{Erf}\left(\frac{bm_2 + a(m_1-r)}{\sqrt{2a}}\right)\right),$$

and

$$\int_{-\infty}^{-r} \exp\left\{-\frac{1}{2}\langle\begin{pmatrix}y-m_1\\-m_2\end{pmatrix}, \begin{pmatrix}a & b\\b & d\end{pmatrix}\begin{pmatrix}y-m_1\\-m_2\end{pmatrix}\rangle\right\} dy$$
$$= \frac{\sqrt{2\pi}}{\sqrt{a}} \exp\left\{-\frac{ad-b^2}{2a}m_2^2\right\}\left(1 - Erf\left(\frac{bm_2 + a(m_1+r)}{\sqrt{2a}}\right)\right).$$

Here Erf denotes the "error function":

$$Erf(z) = \frac{2}{\sqrt{\pi}} \int_0^z e^{-t^2}\, dt\ .$$

Bibliography

[1] K.S. Alexander and V. Sidoravicius, Pinning of polymers and interfaces by random potentials, *Ann. Appl. Pobab. 16 (2006), pp. 636-669.*

[2] L. Alili and R. A. Doney, Wiener-Hopf factorization revisited and some applications, *Stochastics and Stochastics Reports 66 (1999), pp. 87-102.*

[3] S. Asmussen, Applied Probabiltity and Queues, *Springer-Verlag (2003).*

[4] N.H. Bingham, C.M. Goldie and J.L. Teugels, Regular variation, *Cambridge University Press (1989).*

[5] E. Bolthausen, Localization-Delocalization Phenomena for Random Interfaces, *ICM 2002, Vol. III, 1-3 (2002).*

[6] E. Bolthausen, J.-D. Deuschel and O. Zeitouni, Entropic Repulsion of the Lattice Free Field, *Comm. Math. Phys. 170 (1995), pp. 417-443.*

[7] E. Bolthausen, T. Funaki and T. Otobe, Concentration under scaling limits for weakly pinned Gaussian random walks, *Probab. Theory Relat. Fields 143 (2009), pp. 441-480.*

[8] A. Böttcher and B. Silbermann, Introduction to large truncated Toeplitz matrices, *Springer-Verlag (1999).*

[9] R. Bundschuh, M. Lässig and R. Lipowsky, Semiflexible polymers with attractive interactions, *Eur. Phys. J. E 3 (2000), pp. 295-306.*

[10] F. Caravenna and J.-D. Deuschel, Pinning and Wetting Transition for (1+1)-dimensional Fields with Laplacian Interaction, *Ann. Probab. 36 (6) (2008), pp. 2388-2433.*

[11] F. Caravenna and J.-D. Deuschel, Scaling Limits of (1+1)-dimensional Pinning Models with Laplacian Interaction, *Ann. Probab. 37 (3) (2009), pp. 903-945.*

[12] F. Caravenna, G. Giacomin and L. Zambotti, Sharp Asymptotic Behavior for Wetting Models in (1+1)-dimension, *Elect. J. Probab. 11 (2006), pp. 345-362.*

[13] J.-D. Deuschel, G. Giacomin and L. Zambotti, Scaling Limits of Equilibrium Wetting Models in (1+1)-dimension, *Probab. Theory Rel. Fields 132 (2005), pp. 471-500.*

[14] S.P. Eveson, Compactness Criteria for Integral operators in L^∞ and L^1 Spaces, *Proc. Amer. Math. Soc. 123 (12) (1995), pp. 3709-3716.*

[15] W. Feller, An Introduction to Probability Theory and its Applications. Vol. 1, *John Wiley and Sons (1971).*

[16] M.E. Fisher, Walks, Walls, Wetting and Melting, *J. Stat. Phys. 34 (1984), pp. 667-729.*

[17] T. Funaki, A scaling limit for weakly pinned Gaussian random walks, *Proceedings of RIMS Workshop on Stochastic Analysis and Applications B6 (2008), pp. 97-109.*

[18] G. Giacomin, Random Polymer Models, *Imperial College Press (2007).*

[19] P. Gutjahr, J. Kierfeld and R. Lipowsky, Persistence length of semiflexible polymers and bending rigidity renormalization, *Europhys. Lett. 76 (6) (2006), pp. 994-1000.*

[20] Y. Isozaki and N. Yoshida, Weakly Pinned Random Walk on the Wall: Pathwise Descriptions of the Phase Transitions, *Stoch. Proc. Appl. 96 (2001), pp. 261-284.*

[21] T. Kato, Perturbation Theory for Linear Operators, *Springer-Verlag (1976).*

[22] N. Kurt, Entropic repulsion for a class of Gaussian interface models in high dimensions, *Stoch. Proc. Appl. 117 (2007), pp. 23-34.*

[23] N. Kurt, Maximum and entropic repulsion for a Gaussian membrane model in the critical dimension, *Ann. Probab. 37 (2) (2009), pp. 687-725.*

[24] R. Lipowsky, U. Seifert and A. Volmer, Critical behavior of interacting surfaces with tension, *Eur. Phys. J. B 5 (1998), pp. 811-823.*

[25] E. Nummelin, General Irreducible Markov Chains and Non-Negative Operators, *Cambridge University Press (1984).*

[26] P. Rózsa, On the inverse of band matrices, *Integral Equations and Operator Theory 10 (1) (1987), pp. 83-95.*

[27] H. Sakagawa, Entropic repulsion for a Gaussian lattice field with certain finite range interaction, *Journal of Mathematical Physics 44 (7) (2003), pp. 2939-2951.*

[28] H.H. Schaefer, Banach Lattices and Positive Operators, *Springer-Verlag (1974).*

[29] Y. Velenik, Localization and Delocalization of Random Interfaces, *Probability Surveys 3 (2006), pp. 112-169.*

[30] M. Zerner, Quelques propriétés spectrales des opérateurs positifs, *J. Funct. Anal. 72 (1987), pp. 381-417.*

I want morebooks!

Buy your books fast and straightforward online - at one of world's fastest growing online book stores! Environmentally sound due to Print-on-Demand technologies.

Buy your books online at
www.morebooks.shop

Kaufen Sie Ihre Bücher schnell und unkompliziert online – auf einer der am schnellsten wachsenden Buchhandelsplattformen weltweit! Dank Print-On-Demand umwelt- und ressourcenschonend produziert.

Bücher schneller online kaufen
www.morebooks.shop

KS OmniScriptum Publishing
Brivibas gatve 197
LV-1039 Riga, Latvia
Telefax: +371 686 204 55

info@omniscriptum.com
www.omniscriptum.com

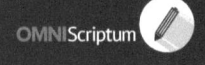

Printed by Books on Demand GmbH, Norderstedt / Germany